零点起飞学编程

零点起飞学 Android 开发

徐 诚 等编著

清华大学出版社
北京

内 容 简 介

本书结合大量实例，由浅入深、循序渐进地介绍了 Android 移动开发技术。本书讲解详细，示例丰富，每一个知识点都配备了具体的示例和运行结果图，可以让读者轻松上手，建立 Android 技术的思想框架，并对 Android 开发过程有个初步了解。本书特意提供了典型习题及教学 PPT 以方便教学。另外，配书光盘中提供了大量的配套教学视频及本书涉及的源代码，便于读者更加高效地学习。

本书共 17 章，分为 2 篇。第 1 篇 Android 开发基础主要介绍了 Android 系统的发展史、基本组件、Android 开发环境的搭建、Android 布局、Android 基本控件、Android 高级控件、Android 辅助功能、Activity 和 Intent、Service 与 BroadcastReceiver、Android 数据存储等。第 2 篇 Android 典型应用与实战重点介绍了 Android 开发中的网络应用、图形应用、多媒体应用、感应器应用、地图服务应用、通信服务及设备控制等，最后通过一个网上购书应用系统的开发，展示了使用 Android 进行实际开发的步骤和流程。

本书适合 Android 移动开发入门与提高人员阅读，也可作为大中专院校及职业院校移动开发类课程的教材。另外，本书也可供从事移动开发的程序员和编程爱好者作为实际工作中的参考书籍。

本书封面贴有清华大学出版社防伪标签，无标签者不得销售。
版权所有，侵权必究。侵权举报电话：010-62782989　13701121933

图书在版编目（CIP）数据

零点起飞学 Android 开发 / 徐诚等编著. —北京：清华大学出版社，2013（2019.12重印）
（零点起飞学编程）
ISBN 978-7-302-32318-1

Ⅰ. ①零… Ⅱ. ①徐… Ⅲ. ①移动终端–应用程序–程序设计 Ⅳ. ①TN929.53

中国版本图书馆 CIP 数据核字（2013）第 092460 号

责任编辑：夏兆彦
封面设计：欧振旭
责任校对：胡伟民
责任印制：李红英

出版发行：清华大学出版社
网　　址：http://www.tup.com.cn, http://www.wqbook.com
地　　址：北京清华大学学研大厦 A 座　　　邮　编：100084
社 总 机：010-62770175　　　　　　　　　　邮　购：010-62786544
投稿与读者服务：010-62776969，c-service@tup.tsinghua.edu.cn
质 量 反 馈：010-62772015，zhiliang@tup.tsinghua.edu.cn

印 装 者：三河市铭诚印务有限公司
经　　销：全国新华书店
开　　本：185mm×260mm　　　印　张：22.5　　　字　数：565 千字
　　　　　附光盘 1 张
版　　次：2013 年 8 月第 1 版　　　　　　　　印　次：2019 年 12 月第 9 次印刷
定　　价：49.80 元

产品编号：051509-01

前　　言

　　Android 是一种基于 Linux 的自由及开放源代码的操作系统，由 Google 公司和开放手机联盟领导及开发的，主要应用于移动设备，比如智能手机和平板电脑，是当前最流行和最热门的移动开发技术之一。无论你是一位 Android 的业余爱好者还是一位程序开发人员，都应该对 Android 系统有一定的了解。

　　随着 Android 应用的普及，国内关于 Android 的图书也如同雨后春笋般出现。这些书多以资深开发者的角度讲述各项技术，对于入门读者而言，由于缺乏相应的从业经验，往往处处碰壁，始终不得要领。基于这个原因，笔者编写了本书，希望能让 Android 入门的新手轻松地进入 Android 移动开发的世界。

　　本书是一本 Android 开发入门读物。考虑新手入门的特点，本书用通俗易懂的语言，有针对性地结合了大量示例，帮助读者掌握每项技术。同时，为了方便读者可以高效而直观地掌握 Android 开发技术，本书提供了全程多媒体教学视频，以辅助读者学习本书的内容。学习完本书后，读者可以熟悉 Android 开发的核心技术，并对 Android 的实际应用开发有个初步的感受，为进一步深入学习打好基础。

本书有何特色

1．门槛低，容易入门

　　相较于市场上的同类图书而言，本书门槛很低。本书只需读者有一定的 Java 程序编写经验即可顺利学习。即使读者没有 Java 开发经验，只要跟着书中的讲解一步步地学习，也能基本掌握书中的知识。

2．语言简洁明了，重点突出，减轻读者阅读负担

　　本书最大的特色就是减轻了读者的阅读负担，以尽可能少的篇幅将 Android 技术的核心知识展示给读者，让读者轻松掌握 Android 技术开发的精髓。

3．示例多，图例多，实用性强

　　为书中的每个知识点都编写了示例进行讲解，便于读者更好地理解和掌握。针对没有接触过 Android 的读者，本书还插入了大量的图片来说明概念，演示操作过程，并给出每个示例的运行效果，让读者切实感受到 Android 技术的强大功能。

4．写作细致，处处为读者着想

　　本书内容编排、概念表述、语法讲解、示例讲解、源代码注释等都很细致。作者讲解时不厌其烦，细致入微，将问题讲解得很清楚，扫清了读者的学习障碍。

5．贯穿大量的开发技巧和注意事项

本书在讲解知识点时使用了大量短小精悍的典型实例，并在这些典型实例讲解中为大家提供了很多开发技巧和注意事项，以使读者迅速提高开发水平。

6．提供配套的多媒体教学视频，体验全新教学课堂

作者专门录制了大量的配套多媒体语音教学视频，以便让读者更加轻松、直观地学习本书内容，提高学习效率。这些视频与本书源代码一起收录于本书配套光盘中。

7．提供教学 PPT，方便老师教学

本书适合大中专院校和职业学校作为职业技能课程的教学用书，所以专门制作了教学PPT，以方便各院校的老师教学时使用。

本书内容安排

第 1 篇　Android 开发基础（第 1~8 章）

本篇主要内容包括 Android 系统的发展史、基本组件简介、Android 开发环境的搭建、Android 常见界面布局、Android 基本控件和高级控件、Android 菜单和对话框、Activity 和 Intent、Service 与 BroadcastReceiver，以及 Android 数据存储。通过本篇的学习，读者可以对 Android 技术有一个大概的了解，并重点掌握 Android 开发的核心技术。

第 2 篇　Android 典型应用与实战（第 9～17 章）

本篇主要内容包括 Android 开发中的网络应用、图形图像应用、多媒体应用、感应器应用、地图服务应用、通信服务及设备控制等方面的知识，最后通过一个网上购书应用系统的开发，展示了使用 Android 进行实际开发的步骤和流程。通过本篇的学习，读者可以掌握 Android 开发中的各种典型应用，并对 Android 的实际应用开发过程有个初步的了解。

本书光盘内容

- 本书配套教学视频；
- 本书实例涉及的源代码。

本书读者对象

- 从未接触过 Android 的初学者；
- 想学习热门开发技术的求职者；
- 初级 Android 开发人员；
- 大中专院校的学生；
- Android 培训班的学员。

本书阅读建议

- ❏ 读者最好有一定的 Java 基础,具备一定的 Java 程序写作能力。
- ❏ 建议没有基础的读者,从前向后顺次阅读,尽量不要跳跃。
- ❏ 建议读者亲自上机动手实践书中的实例和示例,学习效果将会更好。
- ❏ 课后习题都动手做一做,以检查自己对本章内容的掌握程度,如果不能顺利完成,建议重新学习本章的内容。
- ❏ 学习每章内容时,建议读者先仔细阅读书中的讲解,然后再结合本章的教学视频,学习效果会更佳。

本书作者

本书由徐诚主笔编写。其他参与编写的人员有毕梦飞、蔡成立、陈涛、陈晓莉、陈燕、崔栋栋、冯国良、高岱明、黄成、黄会、纪奎秀、江莹、靳华、李凌、李胜君、李雅娟、刘大林、刘惠萍、刘水珍、马月桂、闵智和、秦兰、汪文君、文龙、陈冠军、张昆。

阅读本书的过程中,若有任何疑问,可以发邮件到 book@wanjuanchina.net 或 bookservice2008@163.com,也可以到 www.wanjuanchina.net 的图书论坛上留言,以获得帮助。

编者

目 录

第 1 篇 Android 开发基础

第 1 章 认识 Android（教学视频：38 分钟） ... 2
- 1.1 Android 简介 ... 2
  - 1.1.1 Android 发行版本 ... 2
  - 1.1.2 Android 系统架构 ... 3
  - 1.1.3 Android 组件简介 ... 3
- 1.2 Android 环境搭建 ... 3
  - 1.2.1 下载并安装 JDK ... 4
  - 1.2.2 配置环境变量 ... 4
  - 1.2.3 下载并安装 Eclipse ... 6
  - 1.2.4 下载并安装 AndroidSDK ... 7
  - 1.2.5 安装 Android ADT ... 8
  - 1.2.6 创建运行 AVD ... 9
- 1.3 第一个 Android 程序 ... 11
  - 1.3.1 项目创建 ... 11
  - 1.3.2 项目界面 ... 11
  - 1.3.3 项目运行 ... 13
- 1.4 Android 应用程序结构 ... 14
- 1.5 小结 ... 15
- 1.6 习题 ... 16

第 2 章 Android 常见界面布局（教学视频：43 分钟） ... 18
- 2.1 界面简介 ... 18
- 2.2 相对布局 RelativeLayout ... 19
  - 2.2.1 相对父容器布局 ... 19
  - 2.2.2 相对控件布局 ... 21
- 2.3 线性布局 LinearLayout ... 23
  - 2.3.1 水平线性布局 ... 23
  - 2.3.2 垂直线性布局 ... 25
- 2.4 表格布局 TableLayout ... 25

2.5	网格布局 GridLayout	27
2.6	帧布局 FrameLayout	29
	2.6.1 帧布局	29
	2.6.2 滚动视图 ScrollView	30
	2.6.3 水平滚动视图 HorizontalScrollView	31
2.7	小结	32
2.8	习题	33

第 3 章 Android 常用基本控件（教学视频：42 分钟） 36

3.1	文本控件概述	36
	3.1.1 控件属性	36
	3.1.2 控件使用	37
3.2	文本类控件	37
	3.2.1 TextView	37
	3.2.2 EditText	40
3.3	Button 类控件	41
	3.3.1 Button	42
	3.3.2 ImageButton	43
	3.3.3 ToggleButton	45
	3.3.4 RadioButton	46
	3.3.5 CheckBox	48
3.4	图片控件 ImageView	50
3.5	时钟控件	51
3.6	日期与时间控件	52
	3.6.1 DatePicker	52
	3.6.2 TimePicker	53
3.7	小结	54
3.8	习题	54

第 4 章 Android 高级控件（教学视频：64 分钟） 59

4.1	进度条 ProgressBar	59
4.2	拖动条 SeekBar	61
4.3	自动完成文本控件	63
	4.3.1 使用 AutoCompleteTextView	63
	4.3.2 使用 MultiAutoCompleteTextView	65
4.4	评分条 RatingBar	67
4.5	下拉列表 Spinner	70
4.6	选项卡 TabHost	71
4.7	图片切换控件 ImageSwitcher	74
4.8	列表视图 ListView	77

4.9 网格视图 GridView 80
4.10 小结 82
4.11 习题 82

第5章 Android 菜单和对话框（教学视频：76分钟）... 90
5.1 菜单 Menu 90
　　5.1.1 选项菜单 Options Menu 和子菜单 Submenu 90
　　5.1.2 上下文菜单 Context Menu 94
5.2 对话框 Dialog 96
　　5.2.1 普通对话框 Dialog 96
　　5.2.2 提示对话框 AlertDialog 97
　　5.2.3 进度对话框 ProgressDialog 100
　　5.2.4 日期选择对话框 DatePickerDialog 101
　　5.2.5 时间选择对话框 TimePickerDialog 101
5.3 Android 中的温馨提示 102
　　5.3.1 消息提示条 Toast 102
　　5.3.2 通知 Notification 104
5.4 小结 106
5.5 习题 106

第6章 Activity 和 Intent（教学视频：49分钟）... 111
6.1 Activity 生命周期 111
6.2 单界面程序 112
　　6.2.1 单界面程序启动 112
　　6.2.2 Activity 状态变化 113
　　6.2.3 单界面程序退出 114
6.3 多界面程序 115
　　6.3.1 启动第一个 Activity——主 Activity 115
　　6.3.2 新建第二个 Activity——Two 116
　　6.3.3 启动 Two 118
　　6.3.4 跳转回主 Activity 119
　　6.3.5 BACK 到第二个 Activity 120
6.4 两个 Activity 之间传递数据 121
　　6.4.1 传递数据到目标 Activity 121
　　6.4.2 返回数据到主 Activity 123
6.5 Intent 和 IntentFilter 125
　　6.5.1 意图 Intent 125
　　6.5.2 意图过滤器 IntentFilter 130
6.6 小结 132
6.7 习题 132

第 7 章 Service 与 BroadcastReceiver（教学视频：57 分钟）...... 135

7.1 Service 简介 135
7.1.1 Service 的特点和创建 135
7.1.2 Service 生命周期 136

7.2 Service 操作 137
7.2.1 使用 context.startService()启动 Service 137
7.2.2 使用 context.bindService()启动 Service 140

7.3 Service 通信 142
7.3.1 本地服务通信 142
7.3.2 远程服务通信 145

7.4 系统 Service 149
7.4.1 电话管理器 TelephonyManager 149
7.4.2 短信管理器 SmsManager 152
7.4.3 音频管理器 AudioManager 154
7.4.4 振动器 Vibrator 157

7.5 广播接收者 BroadcastReceiver 158
7.5.1 开发 BroadcastReceiver 158
7.5.2 接收系统广播信息 162

7.6 小结 164
7.7 习题 164

第 8 章 Android 数据存储（教学视频：71 分钟）...... 166

8.1 Android 中存储概要 166
8.2 键值对存储 SharedPreferences 166
8.2.1 SharedPreferences 是什么 166
8.2.2 SharedPreferences 实现数据存储 167

8.3 File 存储 169
8.3.1 File 实现数据读取 169
8.3.2 File 实现 SD 卡中数据的读写 171

8.4 SQLite 数据库存储 174
8.4.1 SQLite 数据库简介 175
8.4.2 数据库编程操作 176
8.4.3 SQLiteOpenHelper 类 178

8.5 数据共享 ContentPrivoder 181
8.5.1 ContentPrivoder 简介 181
8.5.2 ContentProvider 的应用 184

8.6 小结 189
8.7 习题 189

第 2 篇　Android 典型应用与实战

第 9 章　Android 网络应用（教学视频：62 分钟） ……………………………………… 192

- 9.1　Socket 网络通信 …………………………………………………………………… 192
  - 9.1.1　Socket 工作机制 …………………………………………………………… 192
  - 9.1.2　Socket 服务端 ……………………………………………………………… 193
  - 9.1.3　Socket 客户端 ……………………………………………………………… 195
  - 9.1.4　Socket 通信 ………………………………………………………………… 196
- 9.2　HTTP 网络通信 …………………………………………………………………… 197
  - 9.2.1　HTTP 通信方式 …………………………………………………………… 197
  - 9.2.2　HttpURLConnection 开发 ………………………………………………… 198
  - 9.2.3　HttpClient 接口开发 ……………………………………………………… 199
- 9.3　URL 网络通信 ……………………………………………………………………… 205
  - 9.3.1　URL 简介 …………………………………………………………………… 205
  - 9.3.2　URL 通信开发 ……………………………………………………………… 206
- 9.4　WebView 网页开发 ………………………………………………………………… 208
  - 9.4.1　WebView 简介 ……………………………………………………………… 208
  - 9.4.2　WebView 开发 ……………………………………………………………… 209
- 9.5　小结 ………………………………………………………………………………… 212
- 9.6　习题 ………………………………………………………………………………… 212

第 10 章　Android 中图形图像的处理（教学视频：42 分钟） ………………………… 214

- 10.1　Android 中图形图像资源的获取 ………………………………………………… 214
  - 10.1.1　Bitmap 和 Bitmap Factory 类 …………………………………………… 214
  - 10.1.2　获取 assets 文件夹图片资源 …………………………………………… 215
- 10.2　Android 中的动画生成 …………………………………………………………… 217
  - 10.2.1　补间动画 ………………………………………………………………… 217
  - 10.2.2　帧动画 …………………………………………………………………… 221
- 10.3　Android 中图形的绘制 …………………………………………………………… 223
  - 10.3.1　图形绘制类介绍 ………………………………………………………… 223
  - 10.3.2　基本图形的绘制 ………………………………………………………… 224
- 10.4　小结 ………………………………………………………………………………… 226
- 10.5　习题 ………………………………………………………………………………… 226

第 11 章　Android 多媒体应用（教学视频：41 分钟） ………………………………… 228

- 11.1　音乐播放器 ………………………………………………………………………… 228
  - 11.1.1　MediaPlayer 类简介 ……………………………………………………… 228
  - 11.1.2　本地音频文件播放 ……………………………………………………… 229

11.1.3 多个标准音频文件播放 232
11.2 视频播放器 234
    11.2.1 视频相关类简介 234
    11.2.2 视频播放流程 235
11.3 音频与视频的录制 236
    11.3.1 音频录制 236
    11.3.2 视频录制 240
11.4 相机 Camera 243
11.5 小结 245
11.6 习题 245

第 12 章 Android 感应检测——Sensor（教学视频：37 分钟） 248

12.1 Sensor 简介 248
    12.1.1 Sensor 种类 248
    12.1.2 Sensor 开发 248
    12.1.3 Sensor 真机测试 249
    12.1.4 Sensor 信息检测 251
12.2 常用系统传感器 251
    12.2.1 方向传感器 251
    12.2.2 磁场传感器 253
    12.2.3 重力传感器 254
    12.2.4 加速度传感器 255
    12.2.5 光传感器 256
12.3 小结 258
12.4 习题 258

第 13 章 手势识别和无线网络（教学视频：41 分钟） 259

13.1 触摸屏手势 259
    13.1.1 GestureDetector 简介 259
    13.1.2 触摸屏手势应用 260
13.2 输入法手势 262
    13.2.1 Gesture 相关类简介 262
    13.2.2 输入法手势应用 263
13.3 Wi-Fi 265
13.4 蓝牙 Bluetooth 267
13.5 小结 269
13.6 习题 270

第 14 章 Google 地图服务（教学视频：37 分钟） 273

14.1 Google Maps 273
    14.1.1 获取 Map API Key 273

    14.1.2　测试 Google Maps 275
    14.1.3　Google Maps 相关类 276
    14.1.4　Google Maps 应用开发 278
  14.2　Google Street View 284
    14.2.1　Google Street View 服务原理 284
    14.2.2　Google Street View 应用开发 285
  14.3　GPS 定位服务 286
    14.3.1　GPS 相关类简介 287
    14.3.2　GPS 应用开发 288
  14.4　小结 291
  14.5　习题 292

第 15 章　Android 通信服务（教学视频：65 分钟） 294
  15.1　电话控制 294
    15.1.1　拨打电话 294
    15.1.2　过滤电话 298
  15.2　短信控制 300
    15.2.1　发送短信 301
    15.2.2　短信提示 304
    15.2.3　短信群发 307
  15.3　E-mail 控制 310
  15.4　小结 311
  15.5　习题 311

第 16 章　Android 特色应用开发（教学视频：70 分钟） 313
  16.1　手机外观更改和提醒设置 313
    16.1.1　手机壁纸的改变 313
    16.1.2　手机振动的设置 317
    16.1.3　音量调节 320
  16.2　TelephonyManager 的使用 323
  16.3　手机电池电量 327
  16.4　手机闹钟 329
  16.5　小结 332
  16.6　习题 332

第 17 章　Android 应用开发——网上购书（教学视频：43 分钟） 337
  17.1　系统简介 337
    17.1.1　功能概述 337
    17.1.2　开发环境及目标平台 337
  17.2　系统架构 338

17.3 用户登录模块的实现 338
17.4 数据库与数据表的实现 339
17.5 图书浏览选择模块的实现 341
17.6 存储模块的实现 344
17.7 小结 346

第 1 篇　Android 开发基础

- 第 1 章　认识 Android
- 第 2 章　Android 常见界面布局
- 第 3 章　Android 常用基本控件
- 第 4 章　Android 高级控件
- 第 5 章　Android 菜单和对话框
- 第 6 章　Activity 和 Intent
- 第 7 章　Service 与 BroadcastReceiver
- 第 8 章　Android 数据存储

第 1 章 认识 Android

Android 是一种基于 Linux 的自由及开放源代码的操作系统，主要应用于便携设备，如智能手机和平板电脑。2011 年，Android 在全球的智能手机操作系统市场占有份额首次超过塞班系统，跃居全球第一。2012 年，Android 占据全球智能手机操作系统市场 76%的份额，中国市场占有率高达 90%。所以学习 Android 开发具有非常广阔的前景。本章将带领读者步入 Android 的世界，主要介绍 Android 系统的发展、系统框架、开发环境的搭建，以及开发第一个 Android 程序，让读者对 Android 操作系统有一个初步的了解。

1.1 Android 简介

Android 操作系统最初由 Andy Rubin 开发，主要支持手机的操作。2005 年由谷歌收购注资，并组建开放手机联盟进行开发改良，逐渐扩展到平板电脑及其他领域。2008 年 9 月，谷歌正式发布了 Android 1.0 系统，这也是 Android 系统最早的版本。2008 年 10 月，第一部 Android 智能手机发布。

1.1.1 Android 发行版本

从 Android 1.5 版本开始，谷歌开始将 Android 的版本以甜品的名字命名，并设计了不同的 Logo。Android 1.5 是 Cupcake（纸杯蛋糕）、Android 1.6 是 Donut（甜甜圈）、Android 2.1 是 Éclair（松饼）、Android 2.2 是 Froyo（冻酸奶）、Android 2.3 是 Gingerbread（姜饼）、Android 3.0 是 Honeycomb（蜂巢）、Android 4.0 是 Ice Cream Sandwich（冰激凌三明治）、Android 4.1 是 Jelly Bean（果冻豆），如图 1.1 所示。

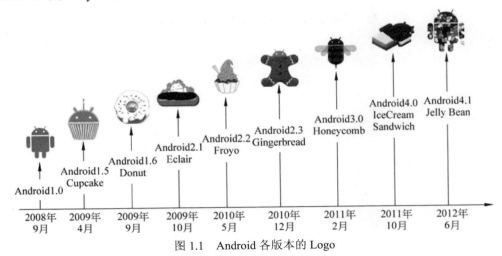

图 1.1　Android 各版本的 Logo

1.1.2　Android 系统架构

Android 的系统架构和它的操作系统一样，采用了分层的架构，如图 1.2 所示。Android 分为 4 个层，从高层到低层分别是应用程序层、应用程序框架层、系统运行库层和 Linux 核心层。

1.1.3　Android 组件简介

Android 常用的组件包括 Activity、Intent、BroadcastReceiver、ContentProvider 和 Service。

图 1.2　Android 系统架构图

1. Activity

在一个 Android 应用中，一个 Activity 就是一个单独的界面。每一个 Activity 被给予一个窗口，在上面可以添加任意控件。窗口通常充满屏幕，但也可以小于屏幕而浮于其他窗口之上，例如对话框。

2. Intent

简单的消息传递框架。使用 Intent 可以在整个系统内广播消息，可以给特定的 Activity 或者服务执行你的行为意图。系统会决定哪个（些）目标来执行适当的行为。例如启动指定的目标组件。

3. BroadcastReceiver

Intent 广播的"消费者"。通过创建和注册一个 BroadcastReceiver，应用程序可以监听符合特定条件的广播的 Intent。BroadcaseReceiver 不执行任何任务，仅仅是接受并响应广播通知的一类组件。

4. ContentProvider

ContentProvider（内容提供器）用来管理和共享应用程序的数据库。在应用程序间，ContentProvider 是共享数据的首选方式。应用程序可以通过 ContentProvider 访问其他应用程序的一些私有数据，这是 Android 提供的一种标准的共享数据的机制。

5. Service

Service 没有用户界面，它会在后台一直运行。主要负责更新数据源和可见的 Activity，以及触发通知。常用来执行一些需要持续运行的处理。例如音乐播放、数据下载等。

1.2　Android 环境搭建

Android 程序由 Java 编程语言开发。因此在开发 Android 程序之前，需要下载并安装开发 Java 程序所需的 JDK，以及运行环境 Eclipse。除此之外，还需要下载并安装 SDK 和

ADT，并配置 JDK 和 SDK 的环境变量。下面我们将依次讲解它们的下载与安装。

1.2.1 下载并安装 JDK

JDK 是 Java 开发工具包。下面以 Java SE 7.0 Development Kits 在 Windows 7 操作系统下的安装为例，逐步搭建 Java 程序的运行环境。

1. 下载 JDK

登录 Oracle 公司的官方网站 http://www.oracle.com/technetwork/java/index.html。在首页单击 DOWNLOADS，跳转到 software Download 页面。单击 Java SE 选项，进入到 Java SE Downloads 页面。单击 Java Platform (JDK) 7u7 进入下载列表。在列表上部，选择 Accept License Agreement 选项，然后选择适合自己计算机的 JDK 版本进行下载。笔者电脑为 Windows 32 位，选择下载 jdk-7u7-windows-i586.exe。下载过程如图 1.3 所示。

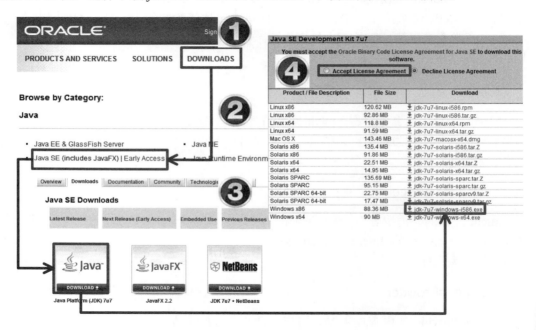

图 1.3　下载 JDK

2. 安装 JDK

双击下载的 JDK 安装包 jdk-7u7-windows-i586.exe，弹出安装向导对话框，安装 JDK 到 C 盘的 Program Files 路径下。安装过程如图 1.4 所示。

至此，JDK 安装完成，位于 C:\Program Files\Java\jdk1.7.0_07 路径下。

1.2.2　配置环境变量

Java 程序运行的环境变量主要包括 Path 和 Classpath。Path 用于指定 JDK 包含的工具程序所在的路径。Classpath 是 Java 程序运行所特需的环境变量，用于指定运行的 Java 程序所需的类的加载路径。

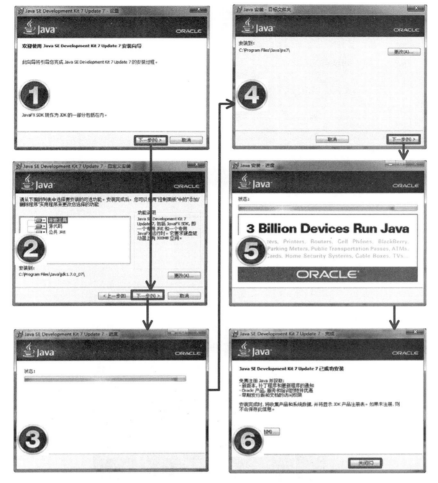

图 1.4 安装 JDK

1. 设置 Path

JDK 包含的工具程序位于 JDK 安装主目录的 bin 目录下，所在的路径为 C:\Program Files\Java\jre1.7.0_07\bin。在 Windows 7 操作系统下，右击桌面上的"计算机"图标，选择"属性"命令，弹出"系统"窗口，选择"高级系统设置"选项，弹出"系统属性"对话框，单击"环境变量"按钮，弹出"环境变量"对话框。

在"环境变量"窗口的"系统变量"栏中选择编辑 Path。在弹出的"编辑系统变量"对话框中的"变量值"文本框后面添加文本";C:\Program Files\Java\jre1.7.0_07\bin"，然后单击"确定"按钮，完成 Path 环境变量设置。设置步骤如图 1.5 所示。

2. 设置 Classpath

在编辑变量时，不像 Path 变量，通常系统没有 Classpath 变量。此时，需要新建一个名为 Classpath 的变量。单击"新建"按钮，弹出"新建系统变量"对话框。在弹出的编辑窗口的"变量名"文本框中输入 Classpath；在"变量值"文本框后面添加文本，配置路径到";C:\Program Files\Java\jre1.7.0_07\lib"。

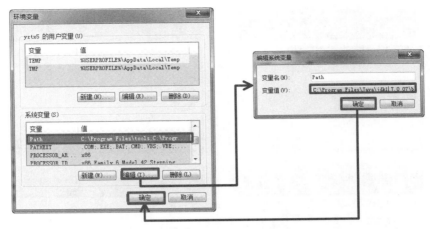

图 1.5　设置 Path 路径

1.2.3　下载并安装 Eclipse

Eclipse 是一个开放源代码的、基于 Java 的可扩展开发平台。从 Eclipse 官方网站 http://www.Eclipse.org 下载目前最新版本的 Eclipse 4.2.0，将其解压到指定位置。

双击解压 eclipse 目录下的 eclipse.exe 文件，弹出 eclipse 启动界面，如图 1.6 所示。

图 1.6　Eclipse 启动界面

在启动过程中，会弹出 WorkSpace Launcher 对话框。用户可以通过单击 Browse…按钮，改变默认的工作空间路径，如图 1.7 所示。

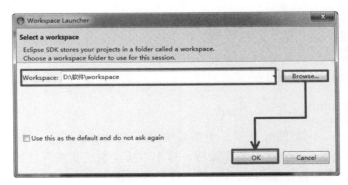

图 1.7　设置工作空间

工作空间是用于存放程序源代码的磁盘空间。在 Eclipse 中编写的程序，都会自动保存在工作空间中。我们可以从工作空间导入已有的项目程序到 Eclipse 中运行。

1.2.4 下载并安装 Android SDK

SDK（Software Development Kit，软件开发工具包），是软件开发工程师为特定的软件包、软件框架、硬件平台、操作系统等建立应用软件时，使用的开发工具的集合。Android SDK 就是用于进行 Android 开发的工具包。下面演示如何在 Windows 7 操作系统下安装 Android SDK。

1. 下载 Android SDK

登录网站 http://developer.android.com/index.html，单击页面下方的 Get the SDK 选项，跳转到 SDK 下载页面，单击 Download the SDK for Windows 按钮下载 SDK，如图 1.8 所示。

图 1.8 下载 SDK

2. 安装配置 SDK

将下载好的 SDK 压缩包解压到 C 盘 Android 文件中，笔者的路径为 C:\Users\yztx5\AppData\Local\Android\android-sdk。然后配置 SDK 的环境变量，在 Path 的"变量值"文本框中添加 C:\Users\yztx5\AppData\Local\Android\android-sdk\platform-tools;。安装过程如图 1.9 所示。

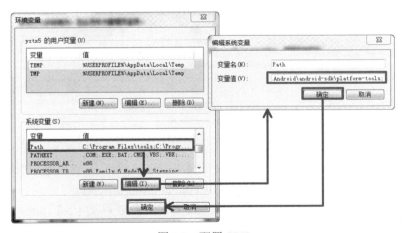

图 1.9 配置 SDK

1.2.5 安装 Android ADT

在 Eclipse 编译环境中，ADT 为 Android 开发提供开发工具的升级或者变更。安装 ADT，首先启动 Eclipse，依次单击菜单 Help|Install|New Software，弹出 Install 对话框，添加可以下载 ADT 的网址进行安装。安装过程如图 1.10 所示。

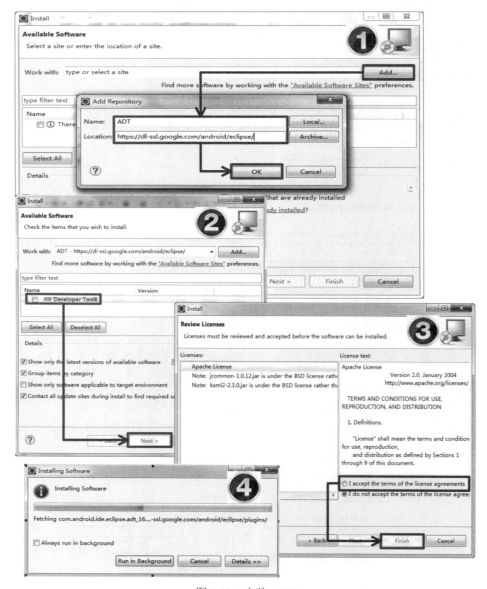

图 1.10　安装 ADT

在 Add Repository 对话框中，在 Name 文本框可以输入任意的名字，Location 文本框中应输入 ADT 的下载地址 https://dl-ssl.google.com/android/eclipse/。

提示：如果输入的地址不能正常下载 ADT，读者可以尝试将 https://dl-ssl.google.com/android/eclipse/改为 http://dl-ssl.google.com/android/eclipse/，或者手动下载 ADT 压缩包。

ADT 安装完成后，会要求用户重启 Eclipse 软件。

1.2.6 创建运行 AVD

AVD（Android Virtual Device）是 Android 运行的虚拟设备。它是 Android 的模拟器识别。AVD 可以代替真机设备，运行我们开发的大量 Android 项目。创建 AVD 的方法有两种：第一种是通过 Eclipse 的 AVD Manager 来创建；第二种是通过命令行创建。下面将分别介绍。

1. Eclipse 的 AVD Manager 创建 AVD

启动 Eclipse，在 Eclipse 的左上角，单击 Opens the Android Virtual Device Manager 按钮，打开 Android 虚拟设备管理器对话框，然后单击 Android Virtual Device Manager 对话框中的 New 按钮新建 AVD。创建步骤如图 1.11 所示。

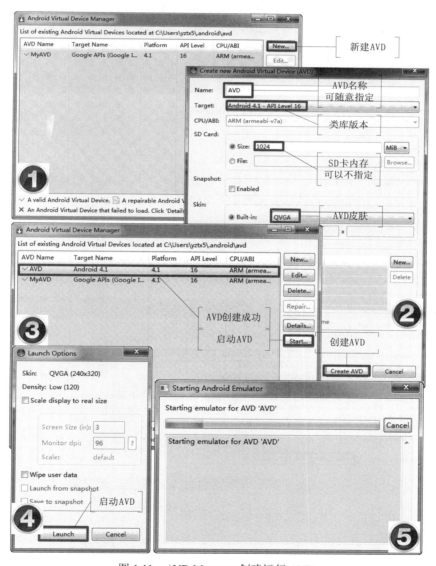

图 1.11 AVD Manager 创建运行 AVD

AVD 成功启动，界面如图 1.12 所示。

图 1.12　AVD 界面图

2. 命令行创建 AVD

打开命令行窗口，在命令行输入如下命令，创建一个名为 avd 的 AVD，target 为 1，位于 D 盘根目录下，皮肤为 QVGA，SD 卡的存储容量为 1024M。

```
android create avd  -n avd  -t 1  -p D:\avd  -s QVGA  -c 1024M  -f
```

提示：-f 表示强制执行创建，会覆盖掉原有的同名 AVD。

在命令行输入如下命令，启动 avd，如图 1.13 所示。

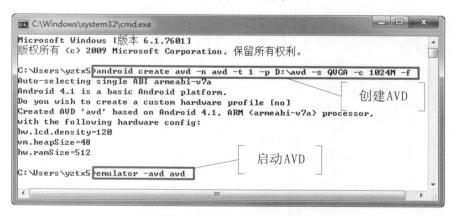

图 1.13　命令行创建启动 AVD

AVD 启动成功，如图 1.14 所示。

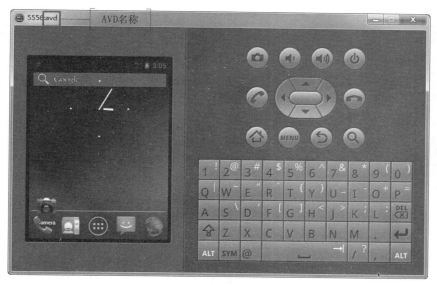

图 1.14 AVD 界面图

1.3 第一个 Android 程序

完成了 Android 运行环境的搭建，下面我们开发第一个 Android 程序——HelloAndroid。

1.3.1 项目创建

按以下步骤新建一个 Android 应用程序，其步骤示意图如图 1.15 所示。

（1）启动 Eclipse，依次选择 File|New|Project 命令，Eclipse 将弹出 New Project 对话框。

（2）单击 Android 下拉箭头，选择 Android Application Project 选项，然后单击 Next 按钮，跳转到 New Android App 对话框。

（3）在 New Android App 对话框中的 Application Name 文本框中输入应用名称 HelloAndroid，然后单击 Next 按钮，切换到 Configure Launcher Icon 界面。

（4）单击 Configure Launcher Icon 界面的 Choose 按钮，可以在弹出的对话框中选择任意图标，作为应用程序图标。然后单击 Next 按钮，切换到 Create Activity 界面。

（5）在 Create Activity 界面中，勾选 Create Activity 的单选框，创建项目的同时，也会自动创建 Activity。然后单击 Next 按钮，切换到 New Blank Activity 界面。

（6）在 New Blank Activity 界面，会自动生成 Activity 的名称和布局文件的名称。读者可以使用默认名称，也可以修改为其他名称。然后单击 Finish 按钮，完成项目创建。

1.3.2 项目界面

项目创建完成后，Eclipse 自动弹出项目布局文件视图界面 Graphical Layout，如图 1.16 所示。可以看到，在手机屏幕中，显示了一行文本"Hello World！"。单击 Graphical Layout 右侧的 activity_main.xml，切换至布局文件代码界面，如图 1.17 所示。

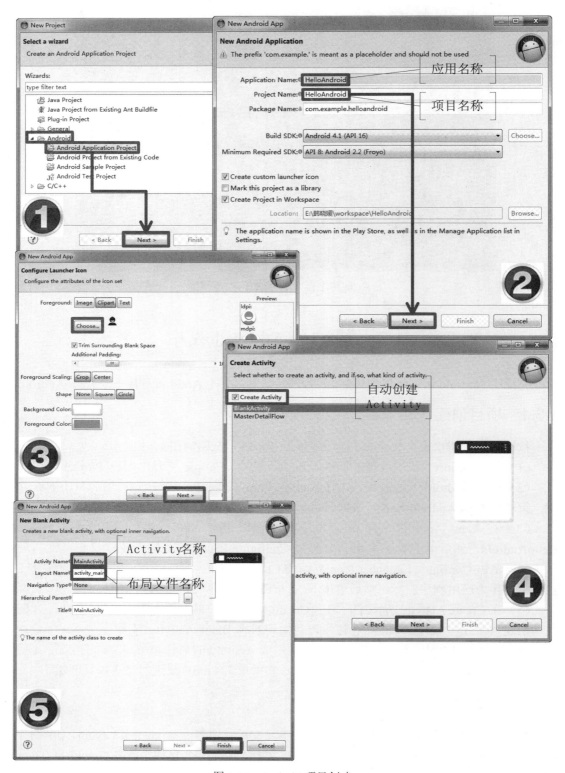

图 1.15 Android 项目创建

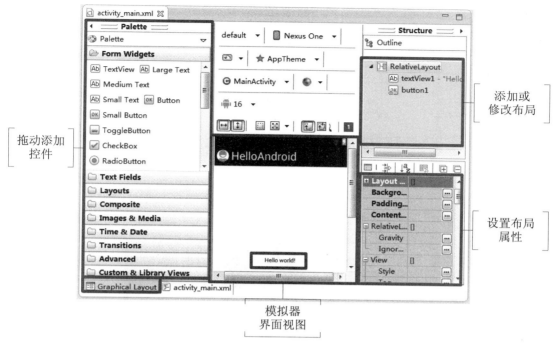

图 1.16 Graphical Layout 界面

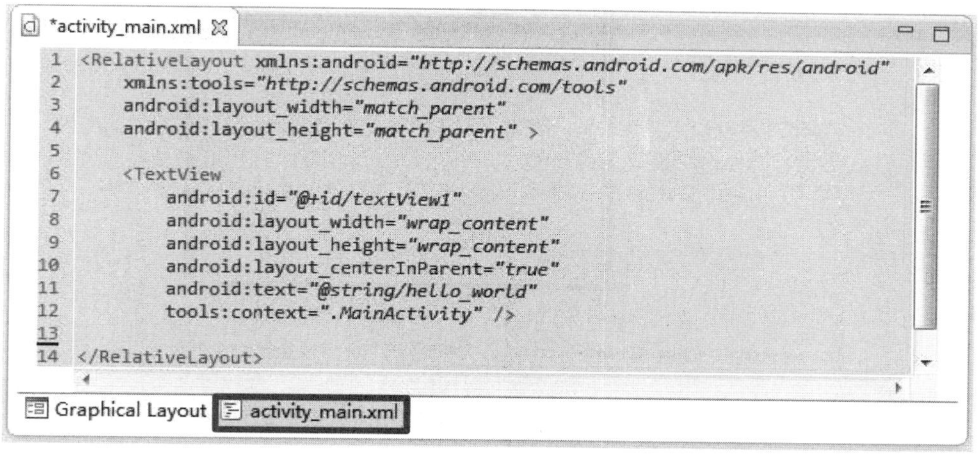

图 1.17 activity_main.xm 界面

在 Eclipse 的 Package Explorer 面板中，进入到 HelloAndroid 项目的 src 目录。单击 com.example.helloandroid 包中的 MainActivity.java 文件，该文件在 Eclipse 右侧打开，如图 1.18 所示。在这里我们可以编写程序逻辑代码。

1.3.3 项目运行

在 Eclipse 的 Package Explorer 面板中，右击 HelloAndroid 项目名称，依次选择 Run As|Android Application 命令，程序开始运行。程序运行成功，如图 1.19 所示。

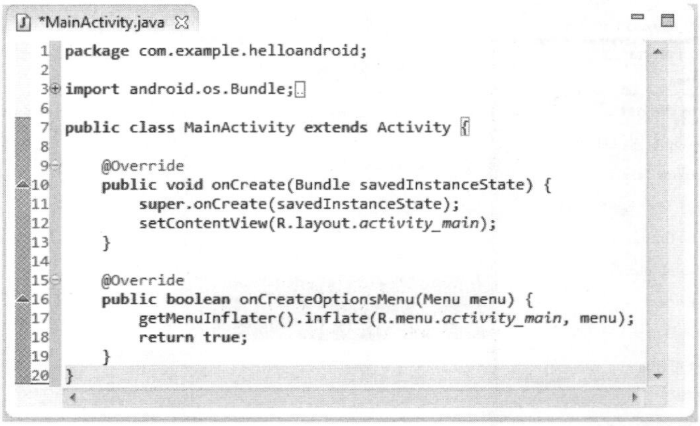

图 1.18　代码编辑界面

图 1.19　项目运行结果图

1.4　Android 应用程序结构

完成了第一个 Android 项目的开发，本节将以此为例，详细介绍 Android 应用程序的结构。如图 1.20 所示为应用程序 HelloAndroid 的结构图。

结构图介绍如下。

- src：存放程序的源代码。
- gen：系统自动生成，无需手动修改。最重要的就是 R.java 文件，保存了程序中所用到的所有控件和资源的 ID。
- assets：存放不进行编译加工的原生文件，这里的资源文件不会在 R.java 自动生成 ID。
- drawable-hdpi：存放高分辨率的资源图片。

第 1 章 认识 Android

图 1.20 HelloAndroid 程序结构图

- drawable-ldpi：存放低分辨率的资源图片。
- drawable-mdpi：存放中等分辨率的资源图片。
- drawable-xhdpi：存放超高分辨率的资源图片，从 Android 2.2 (API Level 8)才开始增加的分类；
- layout：存放项目的布局文件，就是应用程序界面的 XML 文件。
- menu：菜单文件，同样为 XML 格式，在此可以为应用程序添加菜单。
- values：该目录中存放的 XML 文件，定义了各种类型的 key-value 键值对。一般有 dimens、strings、styles、colors、arrays 等。通常程序中用到的尺寸、字符串值、样式、颜色、数组等都在该文件中定义，便于使用和修改。
- AndroidManifest.xml：这是程序的清单文件。应用程序中用到的所有组件，都要在该文件中注册，否则程序无法识别，不能使用。

1.5 小　　结

本章介绍了 Android 系统的发展和 Android 组件，详细讲解了 Android 开发平台的搭建，然后开发了第一个 Android 程序 HelloAndroid，并以该程序为例，介绍了 Android 程序的结构。通过本章的学习，读者应重点掌握 AVD 的创建、Android 项目的创建，以及 Android

程序的结构，对 Android 平台下应用程序的开发步骤有初步的了解。

1.6 习　　题

1. 参考第 1.2 节内容，搭建 Android 程序开发环境，然后在 DOS 窗口，执行 adb 命令，检验 SDK 是否安装成功，效果如图 1.21 所示。

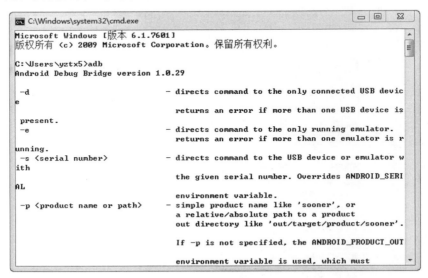

图 1.21　SDK 安装

【分析】本题考查读者对 Android 环境搭建的掌握。首先要确保 Java 运行环境安装配置成功，然后下载安装 Android SDK 并配置路径。如果 adb 命令无法执行，很有可能是环境变量配置不正确。要确保 adb.exe 配置到 path 路径中，注意路径间用分号隔开。第一次安装 ADT，首选安装最新版本，以防出现 Bug，无法开发 Android 程序。

2. 使用 Eclipse 的 AVD Manager 创建 AVD。命名为 avd，设置 SDCard 为 1024M，Skin 为 QVGA。创建完成后，查看该 AVD 属性信息，如图 1.22 所示。启动 avd，效果如图 1.23 所示。

3. 开发第一个 Android 程序。新建项目 FirstAndroid，修改布局文件名称为 activity_first_android。拖动添加一个 TextView 控件，显示文本"我的第一个 Android 程序"，运行程序，效果如图 1.24 所示。

【分析】本题主要考查读者对 Android 程序开发步骤的掌握，并且检验 Android 环境是否搭建成功。可以参考 1.3 节的内容。

【核心代码】本题的核心代码在于布局文件中 TextView 控件的 text 属性设置。代码如下所示。

```
<RelativeLayout
xmlns:android="http://schemas.android.com/apk/res/android"
    xmlns:tools="http://schemas.android.com/tools"
    android:layout_width="match_parent"
```

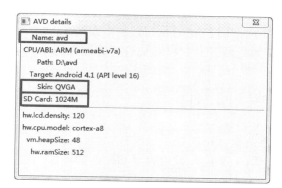

图 1.22 AVD4.1 属性

图 1.23 成功启动 avd

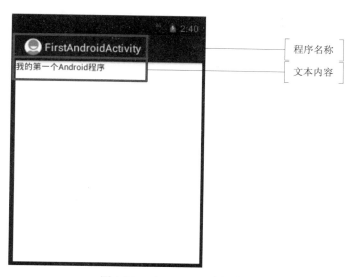

图 1.24 FirstAndroid 界面图

```
        android:layout_height="match_parent" >

    <TextView
        android:id="@+id/textView1"
        android:layout_width="wrap_content"
        android:layout_height="wrap_content"
        android:layout_alignParentLeft="true"
        android:layout_alignParentTop="true"
        android:text="我的第一个Android程序" />

</RelativeLayout>
```

第 2 章 Android 常见界面布局

Android 界面设计被称为布局。一个合理的布局会给用户使用带来更好的体验。Android 中常见的布局包括相对布局 RelativeLayout、线性布局 LinearLayout、表格布局 TableLayout、网格布局 GridLayout 和帧布局 FrameLayout。本章将针对这些布局进行详细介绍。

2.1 界面简介

Android 界面通常由容器和控件构成。容器通常指的是手机屏幕，控件是用于实现功能的图形用户界面元素。布局文件主要是设计 UI 界面，设定容器和控件的属性，规范控件在容器中的显示。本节我们以 HelloAndroid 为例，讲解布局文件。

HelloAndroid 的布局文件 activity_main.xml，位于项目文件的 res 目录下的 layout 子目录中。布局文件采用 XML 格式开发，与程序的逻辑代码分离开来，使程序的开发变得清晰、明了。

布局文件 activity_main.xml 代码如下：

```xml
<RelativeLayout
xmlns:android="http://schemas.android.com/apk/res/android"
    xmlns:tools="http://schemas.android.com/tools"
    <!--设置布局的宽度和高度与父容器匹配-->
    android:layout_width="match_parent"
    android:layout_height="match_parent" >
    <!-- TextView 控件显示文本 HelloWorld -->
    <TextView
        android:layout_width="wrap_content"
        android:layout_height="wrap_content"
        android:layout_alignParentRight="true"
        android:layout_alignParentTop="true"
        android:layout_marginRight="86dp"
        android:layout_marginTop="34dp"
        android:text="@string/hello_world"
        tools:context=".MainActivity" />

</RelativeLayout>
```

属性 layout_width 和 layout_heighth 还有 fill-content 和 wrap-conten 两种参数值，分别表示填充内容和环绕内容。

TextView 的属性设置中，用到了距离单位 dp。除此之外，还有 px 和 sp。

- dp：即 dip（device independent pixels，设备独立像素），不同设备有不同的显示效果，这个和设备硬件有关，一般我们为了支持 WVGA、HVGA 和 QVGA 推荐使用 dp，不依赖像素。

❑ px：即 pixels（像素），不同设备显示效果相同，一般我们用 HVGA 代表 320×480 像素。
❑ sp：即 scaled pixels（放大像素），主要用于字体显示。

如果设置表示长度、高度等属性时可以使用 dp。但如果设置字体大小，需要使用 sp。

2.2 相对布局 RelativeLayout

新建的 Android 应用程序，默认布局文件为相对布局（RelativeLayout）。相对布局分为相对父容器布局和相对控件布局两种，下面将分别介绍。

2.2.1 相对父容器布局

相对父容器布局，主要是针对当前控件边框距父容器（手机屏幕）的四周边框的距离而言，如图 2.1 所示。

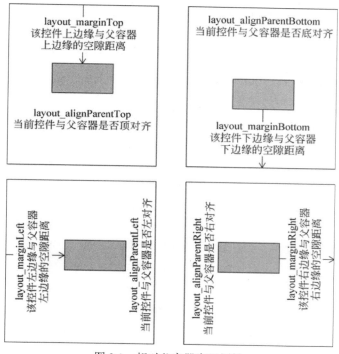

图 2.1 相对父容器布局属性

如果控件距离父容器左右边框相等，可以直接使用 layout_centerHorizontal 设置水平居中，如图 2.2 所示；如果控件距离父容器上下边框相等，可以直接使用 layout_centerVertical 设置垂直居中，如图 2.3 所示；如果控件距离父容器上下边框相等，并且同时距离左右边框也相等，可以使用 layout_centerInParent 设置该控件在整个父容器中居中，如图 2.4 所示。

相对父容器布局语法格式如下：

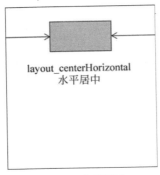 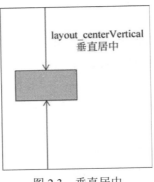 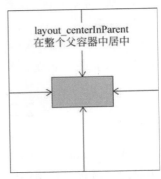

图 2.2　水平居中　　　　图 2.3　垂直居中　　　　图 2.4　父容器居中

```xml
<RelativeLayout
  xmlns:android="http://schemas.android.com/apk/res/android"
  xmlns:tools="http://schemas.android.comtools"
  <!--布局的高度和宽度-->
  android:layout_width="  "
  android:layout_height="  " >

  <Widgets
    ......
    <!-- widgets 在父容器中的居中方式-->
    android:layout_centerHorizontal=" "
    <!--与父容器对齐方式-->
    android:layout_alignParentLeft=" "
    <!--该控件左侧与父容器之间的空隙距离-->
    android:layout_marginLeft="  "

    />
</RelativeLayout>
```

【示例 2-1】　相对父容器布局的使用。在相对布局中添加一个 Button，在父容器中水平居中，Button 顶端与父容器上边框对齐，Button 上边缘距父容器上边缘 64dp。

在 values 目录下的 strings.xml 文件中添加如下代码，显示 Button 文本。

```xml
<string name="button1">Button1</string>
```

布局代码如下：

```xml
<RelativeLayout xmlns:android="http://schemas.android.com/apk/res/android"
  xmlns:tools=http://schemas.android.com/tools
  <!--设置布局的宽度和高度与父容器匹配-->
  android:layout_width="match_parent"
  android:layout_height="match_parent" >

  <Button
    android:id="@+id/button1"
    android:layout_width="wrap_content"
    android:layout_height="wrap_content"
    <!--设置该 button 水平居中与父容器顶对齐-->
    android:layout_alignParentTop="true"
    android:layout_centerHorizontal="true"
    <!--上边缘与父容器相距 64dp-->
    android:layout_marginTop="64dp"
```

```
            android:text="@string/button1" />
</RelativeLayout>
```

注意：Button 的 text 属性的赋值使用了@引用符号。该属性引用了 strings.xml 文件中，name 为 "button1" 的值 "Button1"。所以，Button 显示文本为 Button1。

运行程序，效果如图 2.5 所示。

图 2.5　相对父容器布局

2.2.2　相对控件布局

在相对布局中，未知控件的位置是相对已知控件或是父容器而决定的，如图 2.6 所示。

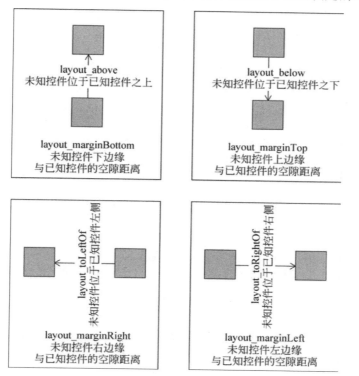

图 2.6　相对控件布局属性

相对控件布局语法格式如下：

```
<RelativeLayout
    xmlns:android="http://schemas.android.com/apk/res/android"
    xmlns:tools="http://schemas.android.com/tools"
<!--布局的宽度和高度-->
    android:layout_width=" "
    android:layout_height=" " >
    <Widgets
    <!--已知控件 ID -->
        android:id="@+id/Widget1"
        ……
        …… />

    <Widgets
        <!--未知控件 ID -->
        android:id="@+id/Widget2"
        android:layout_alignBottom="@+id/Widget1"
        <!--该控件底部与已知控件之间的距离-->
        android:layout_marginBottom=" "
        <!--该控件居于已知控件的右侧-->
        android:layout_toRightOf="@+id/Widget1"/>

</RelativeLayout>
```

【示例 2-2】 相对控件布局的使用。修改【示例 2-1】，再添加一个 Button2，位于 Button1 右下方，并且设置 Button2 上边缘距 Button1 38dp。运行程序，效果如图 2.7 所示。

布局代码如下：

```
<RelativeLayout
xmlns:android="http://schemas.android.com/apk/res/android"
    xmlns:tools="http://schemas.android.com/tools"
    android:layout_width="match_parent"
    android:layout_height="match_parent" >

    <Button
        android:id="@+id/button1"
        android:layout_width="wrap_content"
        android:layout_height="wrap_content"
        android:layout_alignParentTop="true"
        android:layout_centerHorizontal="true"
        android:layout_marginTop="64dp"
        android:text="@string/button1" />

    <Button
        android:id="@+id/button2"
        android:layout_width="wrap_content"
        android:layout_height="wrap_content"
        android:layout_below="@+id/button1"
        android:layout_marginTop="38dp"
        android:layout_toRightOf="@+id/button1"
        android:text="@string/button2" />

</RelativeLayout>
```

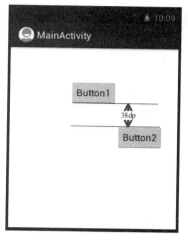

控件	属 性	值
Button1	id	@+id/button1
	layout_width	wrap_content
	layout_heigth	wrap_content
	layout_alignParentTop	ture
	layout_centerHorizonta	ture
	layout_marginTop	64dp
	text	@string/button1
Button2	id	@+id/button2
	layout_width	wrap_content
	layout_heigth	wrap_content
	layout_below	ture
	layout_marginTop	38dp
	layout_toRightOf	@+id/button1
	text	@string/button2

图 2.7 相对控件布局

2.3 线性布局 LinearLayout

线性布局（LinearLayout）是以水平或者垂直的方式来显示界面中添加的控件，因此线性布局可以分为水平线性布局和垂直线性布局两种。下面将分别进行介绍。

2.3.1 水平线性布局

水平线性布局就是从屏幕的左上角开始，将添加的控件以水平的方式排列的布局。在线性布局中使用 android：orientation 属性设置布局的显示方式，horizontal 表示水平显示。水平线性布局语法格式如下：

```
<LinearLayout
    xmlns:android="http://schemas.android.com/apk/res/android"
    xmlns:tools="http://schemas.android.com/tools"
    <!--布局的高度和宽度-->
    android:layout_width=" "
    android:layout_height=" "
    <!--水平显示-->
    android:orientation="horizontal">
    …
    …
</LinearLayout>
```

使用线性布局，需要将新建项目默认的相对布局修改为线性布局。操作方法：打开 activity_main.xml 的 Graphical Layout 视图，在 Outline 面板中，右击 RelativeLayout 分支，从弹出的菜单中选择 Change Layout 命令进行布局修改。修改方法如图 2.8 所示。

【示例 2-3】 水平线性布局的使用。在上述修改好的线性布局文件中添加两个 Button。运行程序，效果如图 2.9 所示。

布局代码如下：

图 2.8 修改布局

```
<LinearLayout xmlns:android="http://schemas.android.com/apk/res/android"
    xmlns:tools="http://schemas.android.com/tools"
    android:layout_width="match_parent"
    android:layout_height="match_parent"
    android:orientation="horizontal">

    ......
    ......

</LinearLayout>
```

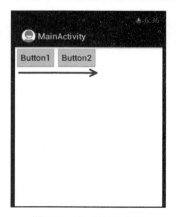

控件	属 性	值
Button1	id	@+id/button1
	layout_width	wrap_content
	layout_heigth	wrap_content
	text	@string/button1
Button2	id	@+id/button1
	layout_width	wrap_content
	layout_heigth	wrap_content
	text	@string/button2

图 2.9 水平线性布局

2.3.2 垂直线性布局

垂直线性布局是指从屏幕的左上角开始,将添加的控件以垂直的方式排列的布局。在线性布局中设置android:orientation属性为vertical,表示布局垂直显示。垂直线性布局语法格式如下:

```
<LinearLayout
    xmlns:android="http://schemas.android.com/apk/res/android"
    xmlns:tools="http://schemas.android.com/tools"
    <!--布局的高度和宽度-->
    android:layout_width=" "
    android:layout_height=" "
    <!--垂直显示-->
    android:orientation="vertical">
    ...
    ...
</LinearLayout>
```

【示例2-4】 垂直线性布局的使用。参考2.3.1节的内容,修改布局为垂直线性布局。然后在修改后的布局文件中添加两个Button。运行程序,效果如图2.10所示。

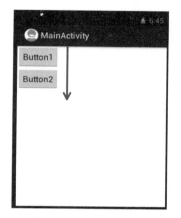

控件	属性	值
Button1	id	@+id/button1
	layout_width	wrap_content
	layout_heigth	wrap_content
	text	@string/button1
Button2	id	@+id/button2
	layout_width	wrap_content
	layout_heigth	wrap_content
	text	@string/button2

图2.10 垂直线性布局

2.4 表格布局 TableLayout

表格布局(TableLayout)将界面划分成多行多列的表格。表格的每行为一个TableRow,每有一个控件添加在TableRow中,就构成一个单元格。每行可以有0个或多个单元格,一个单元格可以跨越多个列。表格布局的相关属性如表2-1所示。

表2-1 表格布局的相关属性

属性名称	属性说明
stretchColumns	指定该列被拉伸,列号从0开始
shrinkColumns	指定该列被收缩,列号从0开始
collapseColumns	指定该列被隐藏,列号从0开始

表格布局语法格式如下：

```xml
<TableLayout
    xmlns:android="http://schemas.android.com/apk/res/android"
    xmlns:tools=http://schemas.android.com/tools
    <!--表格布局的ID -->
    android:id=" "
    <!--布局的高度和宽度-->
    android:layout_width=" "
    android:layout_height=" ">
    <!--指定该列被拉伸-->
    android:stretchColumns=" "
    <!--指定该列被收缩-->
    android:shrinkColumns=" "
    <!--指定该列被隐藏-->
    android:collapseColumns=" "

<TableRow
        <!--当前行的ID -->
        android:id=" "
        <!--当前行的高度和宽度-->
        android:layout_width=" "
        android:layout_height=" ">
        <!--该行中添加第一个控件,形成第一个单元格-->
        <Widgets
            ……
        />
    </TableRow>
</TableLayout>
```

【示例2-5】 表格布局的使用。修改布局为一个两行三列的表格布局，在第一行添加3个Button，在第二行也添加3个Button。然后指定第一列收缩，第二列隐藏，第三列拉伸。运行程序，效果如图2.11所示。

布局代码如下：

```xml
<TableLayout xmlns:android="http://schemas.android.com/apk/res/android"
    xmlns:tools="http://schemas.android.com/tools"
    android:id="@+id/TableLayout1"
    android:layout_width="match_parent"
    android:layout_height="match_parent"
    android:shrinkColumns="0"
    android:stretchColumns="2"
    android:collapseColumns="1"
    >

    <TableRow
        android:id="@+id/tableRow1"
        android:layout_width="wrap_content"
        android:layout_height="wrap_content" >

        <Button
            android:id="@+id/textView1"
            android:layout_width="wrap_content"
            android:layout_height="wrap_content"
            android:text="@string/button1" />

        <Button
            android:id="@+id/textView4"
```

```xml
            android:layout_width="wrap_content"
            android:layout_height="wrap_content"
            android:text="@string/button2" />

        <Button
            android:id="@+id/textView5"
            android:layout_width="wrap_content"
            android:layout_height="wrap_content"
            android:text="@string/button3" />

    </TableRow>

    <TableRow
        android:id="@+id/tableRow2"
        android:layout_width="wrap_content"
        android:layout_height="wrap_content" >

        <Button
            android:id="@+id/textView2"
            android:layout_width="wrap_content"
            android:layout_height="wrap_content"
            android:text="@string/button4" />

        <Button
            android:id="@+id/textView6"
            android:layout_width="wrap_content"
            android:layout_height="wrap_content"
            android:text="@string/button5" />

        <Button
            android:id="@+id/textView7"
            android:layout_width="wrap_content"
            android:layout_height="wrap_content"
            android:text="@string/button6" />
    </TableRow>

</TableLayout>
```

图 2.11 表格布局

控件	属 性	值
TableLayout	stretchColumns	2
	shrinkColumns	0
	collapseColumns	1

2.5 网格布局 GridLayout

　　网格布局（GridLayout）与表格布局相似，用一组无限细的直线将界面分割成行、列、单元，然后指定控件显示的区域和控件在该区域的显示方式。网格布局实现了控件的交错显示，避免使用布局嵌套，更有利于自由编辑布局的开发。网格布局的相关属性如图 2.12 所示。

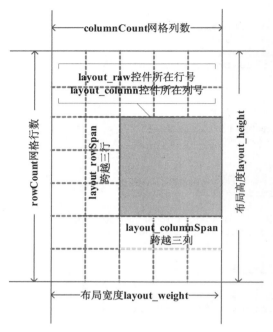

图 2.12　网格布局属性

网格布局语法格式如下：

```
<GridLayout
    xmlns:android="http://schemas.android.com/apk/res/android"
    xmlns:tools=http://schemas.android.com/tools
    <!--网格布局的 ID -->
    android:id=" "
    <!--布局的高度和宽度-->
    android:layout_width=" "
    android:layout_height=" "
    <!--网格布局的行数和列数-->
    android:columnCount=" "
    android:rowCount=" " >

    <Widgets
    <!--控件所在行和列-->
        android:layout_column=" "
        android:layout_row=" "
    <!--控件跨越行数-->
        android:layout_rowSpan=" "
    <!--控件位置-->
        android:layout_gravity=" " />
</GridLayout>
```

【示例 2-6】　网格布局的使用。修改布局为一个三行五列的网格布局。在第一行的第一列单元格添加 Button1；在第二行的第二列添加 Button2，并跨越两列显示；在第三行的第一列添加 Button3，并跨越 3 行显示；在第二行的第一列添加 Space。运行程序，效果如图 2.13 所示。

布局代码如下：

```
<GridLayout xmlns:android="http://schemas.android.com/apk/res/android"
    xmlns:tools="http://schemas.android.com/tools"
    android:id="@+id/GridLayout1"
    android:layout_width="match_parent"
    android:layout_height="match_parent"
    android:columnCount="3"
    android:rowCount="5" >

    ......
    ......

</GridLayout>
```

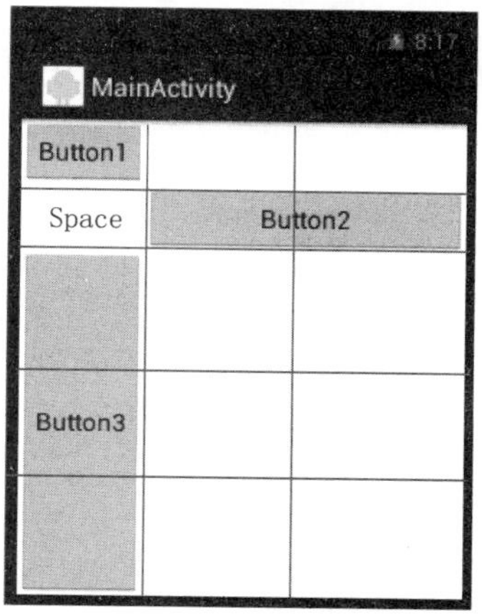

控件	属性	值
Button1	id	@+id/button1
	layout_column	0
	layout_raw	0
	layout_gravity	left
Button2	id	@+id/button2
	layout_column	1
	layout_raw	1
	layout_gravity	fill_horizontal
	layout_columnSpan	2
Button3	id	@+id/button3
	layout_column	0
	layout_raw	2
	layout_gravity	fill_vertical
	layout_rowSpan	3
Space	layout_gravity	fill
	layout_column	0
	layout_raw	1

图 2.13 网格布局

图中的 Space 是一个轻量级的视图子类，用于分隔不同的控件，其中形成一个空白的区域，创建通用布局中组件之间的差距。

2.6 帧布局 FrameLayout

帧布局（FrameLayout）就是为每个加入其中的控件创建一个空白的区域（称为一帧）的布局，每个控件在布局中占据一帧。帧布局中的子类布局 ScrollView 和 HorizontalScrollView，分别支持视图的垂直滚动和水平滚动。当内容过大屏幕无法完全显示时，我们可以滚动视图扩大显示区域。下面将一一介绍。

2.6.1 帧布局

采用帧布局设计界面后，只能在屏幕左上角显示单个控件。如果添加多个控件，则会按顺序叠加在屏幕的左上角重叠显示，但会透明显示出之前控件的文本内容。帧布局语法格式如下：

```
<FrameLayout
    xmlns:android="http://schemas.android.com/apk/res/android"
    xmlns:tools="http://schemas.android.com/tools"
    <!--帧布局 ID -->
    android:id=" "
    <!-- 布局的高度和宽-->
    android:layout_width=" "
    android:layout_height=" " >
    <!--任意控件-->
    <widgets
       ……
        />
    ……
    ……
</FrameLayout>
```

【示例 2-7】 帧布局的使用。修改布局为帧布局，先添加一个较小的 Button1，然后添加一个较大的 Button2。运行程序，效果如图 2.14 所示。从图中可以看出，两个 Button 重叠显示在屏幕左上角，并且透明显示被 Button2 覆盖的 Button1。

布局代码如下：

```
<FrameLayout xmlns:android="http://schemas.android.com/apk/res/android"
    xmlns:tools="http://schemas.android.com/tools"
    android:id="@+id/FrameLayout1"
    android:layout_width="match_parent"
    android:layout_height="match_parent" >
    ……
    ……
</FrameLayout>
```

控件	属性	值
Button1	id	@+id/button1
	layout_width	wrap_content
	layout_heigth	wrap_content
	text	@string/button1
Button2	id	@+id/button2
	layout_width	264dp
	layout_heigth	70dp
	text	@string/button2

图 2.14　帧布局

2.6.2　滚动视图 ScrollView

ScrollView 支持视图垂直滚动，只能拥有一个直接子类。常用的子元素是垂直方向的 LinearLayout，展示一系列的垂直内容。在使用 ScrollView 时，需要将其他布局嵌套在 ScrollView 之内。

ScrollView 语法格式如下：

```
<ScrollView xmlns:android="http://schemas.android.com/apk/res/android"
    xmlns:tools="http://schemas.android.com/tools"
```

```xml
<!--滚动视图 ID -->
    android:id="@+id/scollView1"

    android:layout_width="match_parent"
    android:layout_height="match_parent">

<LinearLayout xmlns:android="http://schemas.android.com/apk/res/android"
    xmlns:tools="http://schemas.android.com/tools"
    android:id="@+id/LinearLayout1"
    android:layout_width="match_parent"
    android:layout_height="wrap_content"
    <!--垂直线性布局-->
    android:orientation="vertical" >

    ……
</LinearLayout>
</ScrollView>
```

【示例 2-8】 ScrollView 的使用。新建项目 ScrollView，修改布局为 ScrollView，然后添加垂直 LinearLayout。在 LinearLayout 中，添加多个 Button，使用 ScrollView 滚动显示视图下方的 Button。运行程序，效果如图 2.15 所示。

布局代码如下：

```xml
<ScrollView xmlns:android="http://schemas.android.com/apk/res/android"
    xmlns:tools="http://schemas.android.com/tools"
    android:id="@+id/scollView1"
    android:layout_width="match_parent"
    android:layout_height="match_parent">

<LinearLayout xmlns:android="http://schemas.android.com/apk/res/android"
    xmlns:tools="http://schemas.android.com/tools"
    android:id="@+id/LinearLayout1"
    android:layout_width="match_parent"
    android:layout_height="wrap_content"
    android:orientation="vertical" >

    ……
    ……
</LinearLayout>
</ScrollView>
```

图 2.15 ScrollView

布局	属性	值
ScrollView	id	@+id/scrollView1
LinearLayout	id	@+id/linearLayout1
	orientation	vertical

2.6.3 水平滚动视图 HorizontalScrollView

HorizontalScrollView 支持视图水平滚动，也只能拥有一个直接子类。常用的子元素是水平方向的 LinearLayout，展示一系列的水平内容。在使用 HorizontalScrollView 时，需要将其他布局嵌套在 ScrollView 之内。

HorizontalScrollView 的语法格式如下：

```
<HorizontalScrollView
    xmlns:android="http://schemas.android.com/apk/res/android"
    xmlns:tools="http://schemas.android.com/tools"
    <!--水平滚动视图 ID -->
    android:id="@+id/HorizontalScrollView1"
    android:layout_width="match_parent"
    android:layout_height="match_parent" >
    <LinearLayout
        xmlns:android="http://schemas.android.com/apk/res/android"
        xmlns:tools="http://schemas.android.com/tools"
        android:id="@+id/LinearLayout1"
        android:layout_width="match_parent"
        android:layout_height="wrap_content"
        <!--水平线性布局-->
        android:orientation="horizontal" >
        ……
    </LinearLayout>
</HorizontalScrollView>
```

【示例 2-9】 HorizontalScrollView 的使用。新建项目 HorizontalScrollView，修改布局为 HorizontalScrollView，然后添加水平 LinearLayout。在 LinearLayout 中，添加多个 Button，使用 ScrollView 滚动显示视图下方的 Button。运行程序，效果如图 2.16 所示。

图 2.16　HorizontalScrollView

布局	属　性	值
HorizontalScrollView	id	@+id/horizontalScrollView
LinearLayout	id	@+id/linearLayout1
	orientation	horizontal

2.7　小　　结

本章主要讲解 Android 的常用布局，包括相对布局、线性布局、表格布局、网格布局和帧布局，以及布局控件 Space。其中，读者需要重点掌握的有相对布局、线性布局、表格布局，以及帧布局中的滚动视图。本章的难点是网格布局和 Space，需要读者多多练习，熟练掌握。

2.8 习　　题

1. 新建项目，采用相对布局，在布局中添加两个 Button 控件。Button1 相对父容器水平居中，Button2 位于 Button1 的左上方。程序运行效果如图 2.17 所示。

【分析】本题目综合考查读者对相对布局的掌握。Button1 是相对父容器布局，Button2 是相对 Button1 布局。可参考 2.2 节的内容。

【核心代码】本题的核心代码如下所示。

```xml
<RelativeLayout
xmlns:android="http://schemas.android.com/apk/res/android"
    xmlns:tools="http://schemas.android.com/tools"
    android:layout_width="match_parent"
    android:layout_height="match_parent" >

    <Button
        android:id="@+id/button1"
        android:layout_width="wrap_content"
        android:layout_height="wrap_content"
        android:layout_alignParentTop="true"
        android:layout_centerHorizontal="true"
        android:layout_marginTop="137dp"
        android:text="@string/button1" />

    <Button
        android:id="@+id/button2"
        android:layout_width="wrap_content"
        android:layout_height="wrap_content"
        android:layout_alignParentTop="true"
        android:layout_marginTop="55dp"
        android:layout_toLeftOf="@+id/button1"
        android:text="@string/button2" />
</RelativeLayout>
```

图 2.17　相对布局

图 2.18　垂直线性布局

2. 新建项目，在 Graphical Layout 视图的 Outline 面板中，修改默认布局为 LinearLayout（vertical），然后添加两个 Button 控件。程序运行效果如图 2.18 所示。

【分析】本题目主要考查读者对修改布局以及线性布局的掌握。可参考 2.3 节的内容。

【核心代码】本题的核心代码如下所示。

```xml
<LinearLayout xmlns:android="http://schemas.android.com/apk/res/android"
    xmlns:tools="http://schemas.android.com/tools"
    android:layout_width="match_parent"
    android:layout_height="match_parent"
    android:orientation="vertical">

    <Button
        android:id="@+id/button1"
        android:layout_width="wrap_content"
        android:layout_height="wrap_content"
        android:text="@string/button1"/>

    <Button
        android:id="@+id/button2"
        android:layout_width="wrap_content"
        android:layout_height="wrap_content"
        android:text="@string/button2"/>

</LinearLayout>
```

3. 新建项目，采用表格布局。在布局中构建一个两行三列的表格，每行添加 3 个按钮。设置第一列隐藏，第二列收缩，第三列拉伸。程序运行效果如图 2.19 所示。

图 2.19　表格布局

【分析】本题目主要考查读者对表格布局的掌握。可参考 2.4 节的内容。

【核心代码】本题的核心代码如下所示。

```xml
<TableLayout xmlns:android="http://schemas.android.com/apk/res/android"
    xmlns:tools="http://schemas.android.com/tools"
    android:id="@+id/TableLayout1"
    android:layout_width="match_parent"
    android:layout_height="match_parent"
    android:shrinkColumns="1"
    android:stretchColumns="2"
    android:collapseColumns="0"
    >
    <TableRow
        android:id="@+id/tableRow1"
        android:layout_width="wrap_content"
        android:layout_height="wrap_content" >
        ……
        ……
    </TableRow>
    <TableRow
```

```
        android:id="@+id/tableRow2"
        android:layout_width="wrap_content"
        android:layout_height="wrap_content" >
        ……
        ……
    </TableRow>
</TableLayout>
```

第 3 章　Android 常用基本控件

我们在进行界面布局时，添加的按钮、文本框、编辑框和图片等，都是 Android 的基本控件。这些控件实现了程序的一些基本功能。本章将针对这类控件进行详细的介绍，使读者掌握基本控件的使用，开发出简单的 Android 程序。

3.1　文本控件概述

Android 系统提供给用户已经封装好的界面控件称为系统控件。系统控件更有利于帮助用户进行快速开发，同时能够使 Android 系统中应用程序的界面保持一致性。

3.1.1　控件属性

Android 支持的基本控件有以下几种，如图 3.1 所示。

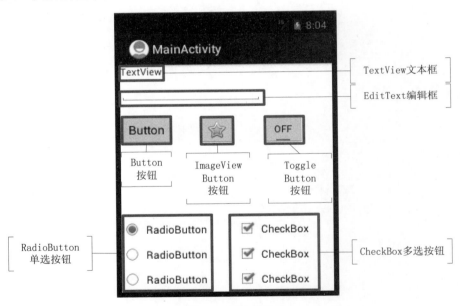

图 3.1　基本控件

注意：由于篇幅有限，图中所列并非 Android 支持的所有基本控件。

Android 的控件，一般是在 res/layout 下的布局文件中声明使用。声明的同时，还要设置控件的属性，控制其在界面中的显示效果。设置控件的属性有两种方法，一种是在布局文件中设置参数，另一种是在代码中调用对应方法实现。控件常用属性及其对应方法如表 3-1 所示。

表 3-1 控件常用属性及其对应方法

属性名称	对应方法	说　明
id	setId(int id)	设置该控件的id
layout_width	setWidth(int pixels)	设置该控件的宽度
layout_height	setHeight(int pixels)	设置该控件的高度

3.1.2　控件使用

在布局文件的 Graphical Layout 视图中有一个 Palette 面板。该面板中包含了 Android 中的所有控件。我们在使用控件时，可以直接拖动所需控件到右侧手机界面，如图 3.2 所示，添加了一个 Button 控件。也可以手动编辑代码添加控件。

图 3.2　添加控件

在布局文件中声明的控件，只负责界面显示。如果要想使用控件实现某些具体功能，就需要在 Activity 中编辑代码实现。实现过程如下：

（1）使用 super.setContentView(R.layout.某布局 layout 文件名)来加载布局文件；
（2）使用 super.findViewById(R.id.控件的 ID)获取控件引用；
（3）使用这个引用对控件进行操作，例如添加监听，设置内容等。

3.2　文本类控件

文本类控件主要用于在界面中显示文本，包含 TextView 和 EditText 两种。下面我们将详细介绍。

3.2.1　TextView

TextView 是 Android 程序开发中最常用的控件之一，它一般使用在需要显示一些信息的时候，它不能输入，只能通过初始化设置或在程序中修改。TextView 常用属性及其对应方法如表 3-2 所示。

表 3-2　TextView 常用属性及对应方法说明

属性名称	对应方法	说明
android:autoLink	setAutoLinkMask(int)	设置是否将指定格式的文本转化为可点击的超链接显示。传入的参数值可取 ALL、EMAIL_ADDRESSES、MAP_ADDRESSES、PHONE_NUMBERS和WEB_URLS
android:height	setHeight(int)	定义TextView的准确高度，以像素为单位
android:width	setWidth(int)	定义TextView的准确宽度，以像素为单位
android:singleLine	setTransformationMethod(TransformationMethod)	设置文本内容只在一行内显示
android:text	setText(CharSequence)	为TextView设置显示的文本内容
android:textColor	setTextColor(ColorStateList)	设置TextView的文本颜色
android:textSize	setTextSize(float)	设置TextView的文本大小
android: textStyle	setTypeface(Typeface)	设置TextView的文本字体
android:ellipsize	setEllipsize(TextUtils.TruncateAt)	如果设置了该属性，当TextView中要显示的内容超过了TextView的长度时，会对内容进行省略，可取的值有start、middle、end和marquee

TextView 文本字体属性示意图如图 3.3 所示。

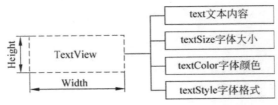

图 3.3　TextView 文本字体属性示意图

TextView 语法格式如下：

```
<TextView
    <!-- TextView 边框包围内容-->
    android:layout_width=" "
    android:layout_height=" "
    <!-- TextView 准确高度宽度-->
    android:width=" "
    android:height=" "
    android:text=" "
    <!--字体大小-->
    android:textSize=" "
    android:textColor=" "
    <!--字体格式-->
    android:textStyle=" "
    <!--文本显示位置-->
    android:gravity=" "
    <!--是否转为可点击的超链接形式-->
    android:autoLink=" "
    <!--是否只在一行内显示全部内容-->
    android:singleLine=" "

    android:ellipsize=" "/>
```

【示例 3-1】　TextView 的使用。新建项目 TextView，在布局中添加三个 TextView。第

一个 TextView 的文本以 web 形式显示 "http://www.google.com"，第二个 TextView 的文本只进行一些字体设置，第三个 TextView 的文本以省略尾部内容显示 26 个英文字母。运行程序，效果如图 3.4 所示。

布局代码如下：

```xml
<TextView
    android:id="@+id/textView1"
    android:layout_width="wrap_content"
    android:layout_height="wrap_content"
    android:text="@string/tv1"
    android:textSize="20sp"
    android:autoLink="web"
    android:singleLine="true"/>

<TextView
    android:id="@+id/textView2"
    android:layout_width="wrap_content"
    android:layout_height="wrap_content"
    android:layout_alignParentLeft="true"
    android:layout_below="@+id/textView1"
    android:layout_marginTop="20dp"
    android:textSize="30sp"
    android:textColor="#0000FF"
    android:textStyle="italic"
    android:text="@string/tv2" />

<TextView
    android:id="@+id/textView3"
    android:layout_width="wrap_content"
    android:layout_height="wrap_content"
    android:layout_alignParentLeft="true"
    android:layout_below="@+id/textView2"
    android:layout_marginTop="20dp"
    android:textSize="30sp"
    android:singleLine="true"
    android:ellipsize="end"
    android:text="@string/tv3" />
```

图 3.4 TextView

控件	属性	值
TextView	id	@+id/textview1
	textSize	20sp
	auto_Link	web
	singleLine	true
	text	http://www.google.com
TextView	id	@+id/textview2
	textSize	30sp
	textColor	#0000ff
	textStyle	italic
	text	TextView
TextView	id	@+id/textview3
	textSize	30sp
	singleLine	true
	ellipsize	end
	text	abcdefghijklmnopqrstuvwxyz

3.2.2 EditText

我们在第一次使用一些应用软件时，常常需要输入用户名和密码进行注册和登录。实现此功能，就需要使用 Android 系统中的编辑框 EditText。EditText 也是一种文本控件，除了 TextView 的一些属性外，EditText 还有一些特有的属性，如表 3-3 所示。

表 3-3 EditText 常用属性及对应方法说明

属 性 名 称	对 应 方 法	说 明
android:lines	setLines(int)	通过设置固定的行数来决定 EditText 的高度
android:maxLines	setMaxLines(int)	设置最大的行数
android:minLines	setMinLines(int)	设置最小的行数
android:inputType	setTransformationMethod(TransformationMethod)	设置文本框中的内容类型，可以是密码、数字、电话号码等类型
android:scrollHorizontally	setHorizontallyScrolling(boolean)	设置文本框是否可以水平滚动
android: capitalize	setKeyListener(KeyListener)	如果设置，自动转换用户输入的内容为大写字母
android: hint	setHint(int)	文本为空时，显示提示信息
android:maxLength	setFilters(InputFilter)	设置最大显示长度

Edittext 属性示意图如图 3.5 所示。

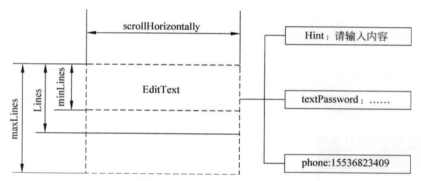

图 3.5 Edittext 属性示意图

EditText 语法格式如下：

```
<EditText
    <!--文本提示内容-->
    android:hint=""
    <!--文本内容显示在固定行中-->
    android:lines=""
    <!--文本最大显示长度-->
    android:maxLength=" "
    <!--文本显示类型-->
    android:inputType=" "

    android:scrollHorizontally=""/>
```

【示例 3-2】 EditText 的使用。新建项目 EditText，在布局文件中添加三个 EditText。

第一个提示输入密码；第二个输入电话号码；第三个输入内容全部转为大写，并限制文本长度。运行程序，效果如图 3.6 所示。

布局代码如下：

```xml
<EditText
    android:id="@+id/EditText1"
    android:layout_width="wrap_content"
    android:layout_height="wrap_content"
    android:layout_alignParentLeft="true"
    android:layout_alignParentTop="true"
    android:password="true"
    android:hint="请输入密码">
</EditText>

<EditText
    android:id="@+id/EditText2"
    android:layout_width="wrap_content"
    android:layout_height="wrap_content"
    android:layout_alignParentLeft="true"
    android:layout_below="@+id/EditText1"
    android:layout_marginTop="26dp"
    android:phoneNumber="true"
    android:lines="1" />

<EditText
    android:id="@+id/EditText3"
    android:layout_width="wrap_content"
    android:layout_height="wrap_content"
    android:layout_alignParentLeft="true"
    android:layout_below="@+id/EditText2"
    android:layout_marginTop="26dp"
    android:maxLength="10"
    android:scrollHorizontally="true"
    android:capitalize="characters" />
```

控件	属性	值
EditText	hint	请输入密码
	password	true
EditText	phoneNumber	true
	lines	1
EditText	maxLength	10
	scrollHorizontally	true
	capitalize	characters

图 3.6　EditText

3.3　Button 类控件

Button 类控件主要包括 Button、ImageButton、ToggleButton、RadioButton 和 CheckBox。下面我们将详细介绍。

3.3.1 Button

Button 是 Android 程序开发过程中，较为常用的一类控件。用户可以通过单击 Button 来触发一系列事件，然后为 Button 注册监听器，来实现 Button 的监听事件。

为 Button 注册监听有两种方法，一种是在布局文件中，为 Button 控件设置 OnCilck 属性，然后在代码中添加一个 public void OnCilck 属性值{}方法；另一种是在代码中绑定匿名监听器，并且重写 onClick 方法。下面我们通过例子来演示为 Button 注册监听。

【示例 3-3】 新建项目 Button，在布局中添加 Button1 和 Button2。在 Activity 中编辑代码为 Button1 注册监听，单击 Button1，修改界面标题"Button1 注册成功"；在布局文件中为 Button2 设置 OnClick 属性值注册监听，单击 Button2，修改界面标题"Button2 注册成功"。

布局文件代码：

```xml
<Button
    android:id="@+id/button1"
    android:layout_width="wrap_content"
    android:layout_height="wrap_content"
    android:text="@string/button1"
    />

<Button
    android:id="@+id/button2"
    android:layout_width="wrap_content"
    android:layout_height="wrap_content"
    android:layout_alignParentLeft="true"
    android:layout_marginTop="60dp"
    <!--设置 OnClick 属性-->
    android:onClick="click"
    android:text="@string/button2" />
```

逻辑代码：

```java
public class MainActivity extends Activity {
    //声明 Button1、Button2
    Button button1,button2;
    @Override
    public void onCreate(Bundle savedInstanceState) {
        super.onCreate(savedInstanceState);
        //加载布局文件
        setContentView(R.layout.activity_main);
        //获取 Button1、Button2 引用
        button1 = (Button)findViewById(R.id.button1);
        button2 = (Button)findViewById(R.id.button2);
        //为 Button1 注册监听
        button1 .setOnClickListener(new OnClickListener() {
            public void onClick(View v) {
                // TODO Auto-generated method stub
                setTitle("Button1 注册成功");
            }
        });
    }

        //为 Button2 注册监听,方法名为 OnClick 属性值
        public void click(View v) {
```

```
            setTitle("Button2注册成功");
        }
}
```

注意：Button 控件的 OnClick 属性，其参数值为在代码中添加的对应方法名，因此在设置该参数值时，需注意命名规范。

程序执行过程如图 3.7 所示。

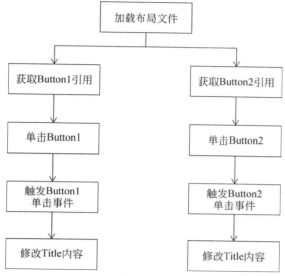

图 3.7　程序执行流程图

运行程序，效果如图 3.8 所示。

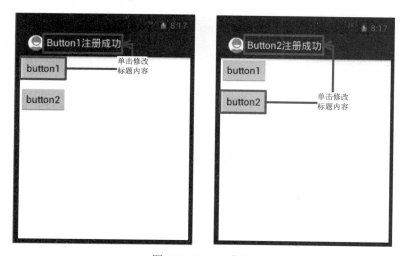

图 3.8　Button 监听

3.3.2　ImageButton

ImageButton（图片按钮）也是一种 Button。它与 Button 控件类似，只是在设置图片时有些区别。ImageButton 控件中，设置按钮显示的图片可以通过 android:src 属性，也可以通过 setImageResource(int)方法来设置。

ImageButton 语法格式如下：

```xml
<ImageButton
    <!-- ImageButton 按钮的 ID -->
    android:id=" "
    <!-- ImageButton 宽度和高度-->
    android:layout_width=" "
    android:layout_height=" "
    <!-- ImageButton 背景图片-->
    android:src=" " />
```

【示例 3-4】 ImageButton 的使用。新建项目 ImageButton，添加两个 ImageButton 控件。第一个使用 drawable 中的图片资源作为按钮背景，第二个使用系统提供的图片作为按钮背景。运行程序，效果如图 3.8 所示。

布局代码如下：

```xml
<ImageButton
    android:id="@+id/imageButton1"
    android:layout_width="wrap_content"
    android:layout_height="wrap_content"
    android:layout_alignParentLeft="true"
    android:layout_alignParentTop="true"
    android:src="@drawable/paint" />

<ImageButton
    android:id="@+id/imageButton2"
    android:layout_width="wrap_content"
    android:layout_height="wrap_content"
    android:layout_alignParentLeft="true"
    android:layout_below="@+id/imageButton1"
    android:layout_marginTop="42dp"
    android:src="@android:drawable/btn_minus" />
```

控件	属性	值
ImageButton	src	@drawable/paint
ImageButton	src	@android:drawable/btn_minus

图 3.8 ImageButton

注意：在设置 src 属性时，加载 Drawable 对象，参数值则为@drawable/对象名；加载系统提供的资源图片，参数值则为@android:drawable/图片名。

【示例 3-5】 下面演示一个单击 ImageButton，改变其背景图片的案例。

首先，在 res/drawable-mdpi 目录下新建一个 myselector.xml，在其中输入如下代码：

```xml
<?xml version="1.0" encoding="utf-8"?>
<selector xmlns:android="http://schemas.android.com/apk/res/android">
    <!--未点击时显示背景-->
    <item android:state_pressed="false"
        android:drawable="@drawable/ic_action_search" />
    <!--点击时显示背景-->
    <item android:state_pressed="true"
        android:drawable="@drawable/ic_launcher" />
</selector>
```

然后，设置布局文件中，ImageButton 控件的 src 属性参数为 myselector.xml 的引用：

```
android:src="@drawable/myselector"
```

selector 是 Android 控件的背景选择器，采用 XML 文件格式。我们可以通过设置 item 项中的以下属性，然后引用图片改变 ImageButton 显示背景。

- android:state_selected：选中；
- android:state_focused：获得焦点；
- android:state_pressed：点击；
- android:state_enabled：设置是否响应事件。

运行程序，效果如图 3.9 所示。

图 3.9　单击改变 ImageButton 背景图片

3.3.3　ToggleButton

ToggleButton（开关按钮）是 Android 系统中比较简单的一个组件，它带有亮度指示，具有选中和未选中两种状态（默认为未选中状态），并且需要为不同的状态设置不同的显示文本。ToggleButton 常用属性及对应方法如表 3-4 所示。

表 3-4　ToggleButton 常用属性及对应方法说明

属性名称	对应方法	说明
android:disabledAlpha		设置按钮在禁用时的透明度，属性值必须为浮点型
android:textoff	setTextOff(CharSequence textOff)	未选中时按钮的文本
android:texton	setTextOn(CharSequence textOn)	选中时按钮的文本

ToggleButton 语法格式如下：

```
<ToggleButton
    <!-- ToggleButton 按钮的 ID -->
    android:id=" "
    <!-- ToggleButton 被选中时显示的文本内容-->
    android:textOn=" "
    <!-- ToggleButton 未被选中时显示的文本内容-->
    android:textOff=" "/>
```

【示例 3-6】 ToggleButton 的使用。新建项目 ToggleButton，在布局文件中添加一个 ToggleButton 控件。设置其被选中时显示"开"，未被选中时显示"关"。运行程序，效果如图 3.10 所示。

布局代码如下：

```
<ToggleButton
    android:id="@+id/toggleButton1"
    android:layout_width="1500dp"
    android:layout_height="80dp"
    android:layout_alignParentLeft="true"
    android:layout_alignParentTop="true"
    android:textOn="开"
    android:textOff="关"/>
```

控件	属性	值
ToggleButton	textOn	开
	textOff	关

图 3.10　ToggleButton

3.3.4　RadioButton

RadioButton（单选按钮）在 Android 平台上也比较常用，比如一些选择项会用到单选按钮。它是一种单个圆形单选框双状态的按钮，可以选择或不选择。在 RadioButton 没有被选中时，用户通过单击来选中它。但是，在选中后，无法通过单击取消选中。

单选按钮由 RadioButton 和 RadioGroup 两部分组成。RadioGroup 是单选组合框，用于将 RadioButton 框起来。在多个 RadioButton 被 RadioGroup 包含的情况下，同一时刻只可以选择一个 RadioButton，并用 setOnCheckedChangeListener 来对 RadioGroup 进行监听。RadioButton 语法格式如下：

```
<RadioGroup
    <!-- RadioGroup 单选组合框的 ID -->
```

```
        android:id=" "
    <!-- RadioButton 排列方式-->
        android:orientation=" " >
    <RadioButton
    <!-- RadioButton 单选按钮的 ID -->
        android:id=" "
    <!-- RadioButton 文本内容-->
        android:text=" " />
    ......
</RadioGroup>
```

【示例 3-7】 RadioButton 的使用。新建项目 RadioButton，在布局文件中添加一个 TextView 显示"请选择："；添加一个 RadioGroup 控件，设置 RadioButton 以垂直方式排列；在 RadioGroup 控件中添加两个 RadioButton 控件，分别显示"火车"和"飞机"；再添加一个 TextView 显示"您选择的是："。在逻辑代码部分编辑代码，当选中不同选项时，在第二个 TextView 后追加显示选项内容。运行程序，效果如图 3.11 所示。

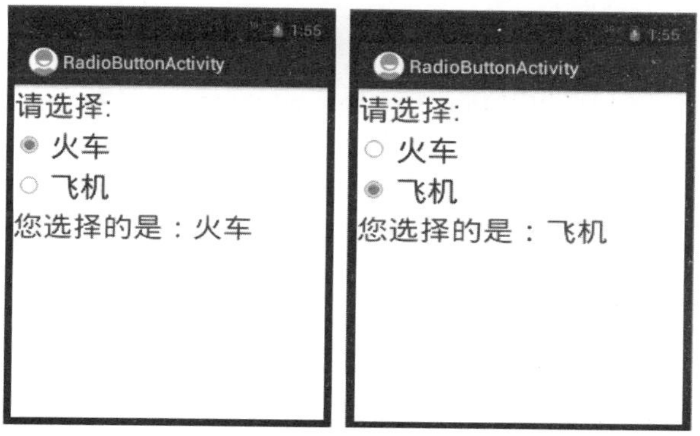

控件	属 性	值
TextView	id	@+id/tv1
	text	请选择:
RadioGroup	id	@+id/rg1
	orientation	vertical
Radiobutton	id	@+id/rb1
	text	火车
Radiobutton	id	@+id/rb2
	text	飞机
TextView	id	@+id/tv2
	text	您选择的是:

图 3.11　RadioButton

布局代码如下：

```
<TextView
    android:id="@+id/tv1"
    android:layout_width="wrap_content"
    android:layout_height="wrap_content"
    android:textSize="30sp"
    android:text="请选择:" />
<RadioGroup
    android:id="@+id/rg1"
    android:layout_width="wrap_content"
    android:layout_height="wrap_content"
    android:orientation="vertical" >
    <RadioButton
        android:id="@+id/rb1"
        android:layout_width="wrap_content"
        android:layout_height="wrap_content"
        android:textSize="30sp"
        android:text="火车" />
    <RadioButton
        android:id="@+id/rb2"
        android:layout_width="wrap_content"
        android:layout_height="wrap_content"
        android:textSize="30sp"
        android:text="飞机" />
```

```xml
</RadioGroup>
<TextView
    android:id="@+id/tv2"
    android:layout_width="wrap_content"
    android:layout_height="wrap_content"
    android:textSize="30sp"
    android:text="您选择的是：" />
```

关键逻辑代码：

```java
//为 RadioGroup 注册监听
radioGroup.setOnCheckedChangeListener(new RadioGroup.OnCheckedChangeListener(){
            public void onCheckedChanged(RadioGroup group, int checkedId){
            // TODO Auto-generated method stub
            //通过 id 判断第一个 RadioButton 被选中
            if (checkedId == R.id.rb1) {
                //显示第一个 RadioButton 内容
                textView.setText("您选择的是：" + radioButton1.getText());
            //第二个 RadioButton 被选中
            }else {
                //显示第二个 RadioButton 内容
                textView.setText("您选择的是：" + radioButton2.getText());
            }
        }
    });
```

3.3.5 CheckBox

CheckBox（复选按钮），顾名思义是一种可以进行多选的按钮，默认以矩形表示。与 RadioButton 相同，它也有选中或者不选中双状态。我们可以先在布局文件中定义多选按钮，然后对每一个多选按钮进行事件监听 setOnCheckedChangeListener，通过 isChecked 来判断选项是否被选中，做出相应的事件响应。CheckBox 语法格式如下：

```xml
<CheckBox
    <!-- CheckBox 复选按钮 ID-->
    android:id=" "
    <!-- CheckBox 文本内容-->
    android:text=" " />
```

【示例 3-8】CheckBox 的使用。新建项目 CheckBox，在布局文件中添加一个 TextView 显示"请选择"；添加三个 CheckBox 控件，分别显示"火车"、"飞机"和"轮船"；再添加一个 TextView 显示"您选择的是:"。在逻辑代码部分编辑代码，当选中不同选项时，在第二个 TextView 后追加显示选项内容。运行程序，效果如图 3.12 所示。

布局代码如下：

```xml
<TextView
    android:id="@+id/textView1"
    android:layout_width="121dp"
    android:layout_height="wrap_content"
    android:textSize="25sp"
    android:text="请选择" />

<CheckBox
    android:id="@+id/checkBox1"
```

```xml
    android:layout_width="wrap_content"
    android:layout_height="wrap_content"
    android:textSize="25sp"
    android:text="火车" />

<CheckBox
    android:id="@+id/checkBox2"
    android:layout_width="wrap_content"
    android:layout_height="wrap_content"
    android:textSize="25sp"
    android:text="飞机" />

<CheckBox
    android:id="@+id/checkBox3"
    android:layout_width="wrap_content"
    android:layout_height="wrap_content"
    android:textSize="25sp"
    android:text="轮船" />

<TextView
    android:id="@+id/textview2"
    android:layout_width="wrap_content"
    android:layout_height="wrap_content"
    android:textSize="25sp"
    android:text="您选择的是：" />
```

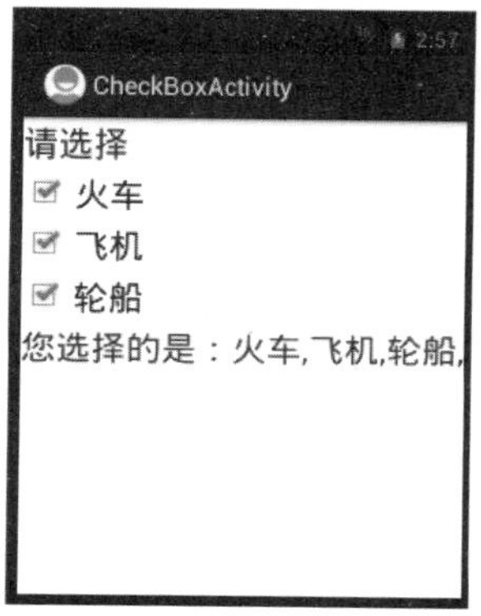

图 3.12 CheckBox

控件	属性	值
TextView	id	@+id/textview1
	text	请选择
CheckBox	id	@+id/checkbox1
	text	火车
CheckBox	id	@+id/checkbox2
	text	飞机
CheckBox	id	@+id/checkbox3
	text	轮船
TextView	id	@+id/textview2
	text	您选择的是:

关键逻辑代码：

```java
//为第一个 CheckBox 注册监听
checkBox1.setOnCheckedChangeListener(new OnCheckedChangeListener() {
    public void onCheckedChanged(CompoundButton buttonView,
        boolean isChecked){
        //如果第一个 CheckBox 被选中
        if (isChecked == true) {
            //显示第一个 CheckBox 内容
            textView.append(checkBox1 .getText() + ",");
        }
    }
});
//为第二个 CheckBox 注册监听
checkBox2.setOnCheckedChangeListener(new OnCheckedChangeListener()
{
    //如果第二个 CheckBox 被选中
```

```
            public    void   onCheckedChanged(CompoundButton    buttonView,
boolean isChecked) {
            if (isChecked == true) {
                //显示第二个 CheckBox 内容
                textView.append(checkBox2 .getText() + ",");
            }
        }
    });
    //为第三个 CheckBox 注册监听
    checkBox3.setOnCheckedChangeListener(new OnCheckedChangeListener()
{
        public void onCheckedChanged(CompoundButton buttonView,
        boolean isChecked) {
        //如果第三个 CheckBox 被选中
            if (isChecked == true) {
                //显示第三个 CheckBox 内容
                textView.append(checkBox3 .getText() + ",");
            }
        }
    });
```

3.4 图片控件 ImageView

ImageView 是一个图片控件，负责显示图片。图片的来源可以是系统提供的资源文件，也可以是 Drawable 对象。ImageView 常用的属性及其对应方法如表 3-5 所示。

表 3-5 ImageView 常用属性及对应方法说明

属 性 名 称	对 应 方 法	说 明
android:adjustViewBounds	setAdjustViewBounds(boolean)	设置是否需要 ImageView 调整自己的边界来保证所显示的图片的长宽比例
android:maxHeight	setMaxHeight(int)	ImageView 的最大高度，可选
android:maxWidth	setMaxWidth(int)	ImageView 的最大宽度，可选
android:scaleType	setScaleType(ImageView.ScaleType)	控制图片应如何调整或移动来适合 ImageView 的尺寸
android:src	setImageResource(int)	设置 ImageView 要显示的图片

ImageView 语法格式如下：

```
<ImageView
    <!-- ImageView 图片控件 ID-->
    android:id=" "
    <!--是否保持长宽比-->
    android:adjustViewBounds=" "
    <!-- ImageView 最大高度和最大宽度-->
    android:maxHeight=" "
    android:maxWidth=" "
    <!--是否调整图片适应 ImageView-->
    android:scaleType=" "
    android:src=" " />
```

【示例 3-9】 ImageView 的使用。新建项目 ImageView，在布局文件中添加两个

ImageView，第一个显示系统图片，第二个显示 drawable 图片。运行程序，效果如图 3.13 所示。

布局代码如下：

```xml
<ImageView
    android:id="@+id/imageView1"
    android:layout_width="wrap_content"
    android:layout_height="wrap_content"
    android:layout_alignParentLeft="true"
    android:layout_alignParentTop="true"
    android:src="@android:drawable/btn_star" />

<ImageView
    android:id="@+id/imageView2"
    android:layout_width="wrap_content"
    android:layout_height="wrap_content"
    android:layout_alignParentLeft="true"
    android:layout_below="@+id/imageView1"
    android:layout_marginTop="74dp"
    android:adjustViewBounds="true"
    android:maxHeight="300dp"
    android:maxWidth="300dp"
    android:scaleType="fitXY"
    android:src="@drawable/paint" />
```

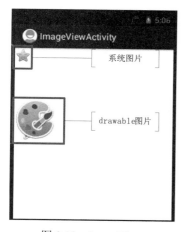

控件	属性	值
ImageView	id	@+id/imageView1
	src	@android:drawable/btn_star
ImageView	id	@+id/imageView2
	src	@drawable/paint
	adjustViewBounds	true
	scaleType	fitXY
	maxHeight	300dp
	maxWidth	300dp

图 3.13　ImageView

3.5　时 钟 控 件

时钟控件包括 AnalogClock 和 DigtialClock，这两种控件都负责显示时间。不同的是，AnalogClock 是模拟时钟，只显示时针和分针；而 DigtialClock 显示数字时钟，可精确到秒。两者可以结合使用，能更准确的表达时间。

【示例 3-10】 结合使用 AnalogClock 和 DigtialClock。新建项目 Clock，在布局文件中添加一个 AnalogClock 控件和一个 DigtialClock 控件，显示系统时间。运行程序，效果如图 3.14 所示。

图 3.14　AnalogClock 和 DigtialClock 结合使用

3.6　日期与时间控件

Android 为用户提供了显示日期与时间的控件 DatePicker 和 TimePicker，下面我们将详细介绍。

3.6.1　DatePicker

日期选择控件（DatePicker）主要的功能向用户提供包含了年、月、日的日期数据，并允许用户对其进行选择。DatePicker 相关属性如表 3-6 所示。

表 3-6　DatePicker 相关属性

属 性 名 称	属 性 说 明
calendarViewShown	是否显示日历视图
maxDate	日历视图显示的最大日期，格式为 mm/dd/yyyy
minDate	日历视图显示的最小日期，格式为 mm/dd/yyyy
spinnersShown	是否显示微调控件

DatePicker 语法格式如下：

```
<DatePicker
        <!-- DatePicker ID-- >
        android:id=" "
        <!--是否显示日历视图-->
        android:calendarViewShown=" "
        <!--日历视图显示的最小日期和最大日期,格式为 mm/dd/yyyy-- >
        android:minDate=" "
        android:maxDate=" "
        <!--是否调整图片适应ImageView-->
        android:spinnersShown=" " />
```

【示例 3-11】DatePicker 的使用。新建项目 DatePicker，在布局文件中添加一个 DatePicker 显示系统日期。设置其显示日历视图和微调控件，并设定日历视图显示的最大日期和最小日期。运行程序，效果如图 3.15 所示。使用微调控件，可以修改日期。

布局代码如下：

```
<DatePicker
    android:id="@+id/datePicker1"
    android:layout_width="wrap_content"
    android:layout_height="wrap_content"
    android:layout_alignParentLeft="true"
    android:layout_alignParentTop="true"
    android:minDate="1-1-1970"
    android:maxDate="12-31-2040"/>
```

控件	属性	值
DatePicker	id	@+id/datePicker1
	calendarViewShown	true
	maxDate	12-31-2040
	minDate	1-1-1970
	spinnersShown	true

图 3.15 DatePicker

如果将上述布局文件中 DatePicker 的 android:spinnersShown 属性设置为 false，就只显示日历视图，如图 3.16 所示。

图 3.16 只显示日历视图

3.6.2 TimePicker

时间选择控件（TimePicker）向用户显示一天中的时间（可以为 24 小时制，也可以为 AM/PM 制），并允许用户进行修改。

【示例 3-12】 TimePicker 的使用。新建项目 TimePicker，在布局文件中添加一个 TimePicker，以 AM/PM 制显示系统时间。运行程序，效果如图 3.17 所示。

【示例 3-13】 在 TimePickerActivity 中添加代码，使用 TimePicker 以 24 小时制显示

系统时间。运行程序,效果如图 3.18 所示。

图 3.17　AM/PM 制 TimePicker

图 3.18　24 小时制 TimePicker

关键代码如下:

```
<!--设置 TimePicker 为 24 小时制-- >
timePicker.setIs24HourView(true);
<!--设置为 14 时-- >
timePicker.setCurrentHour(14);
<!--设置为 40 分-->
timePicker.setCurrentMinute(40);
```

3.7　小　　结

本章主要介绍了 Android 中一些常用的、比较简单的控件。其中,Button 类控件需要注册监听,实现具体功能,需要读者认真学习掌握。掌握这些控件的用法,并结合上一章的布局知识,就能够开发出简单的用户界面。

3.8　习　　题

1. 新建项目 EditText,在布局中添加两个 EditText。第一个显示为密码格式,在未输入密码时,显示文本"请输入密码";第二个显示电话号码,输入电话号码时,界面弹出拨号盘。程序运行效果如图 3.19 所示。

图 3.19　EditText

【分析】本题目主要考查读者对 EditText 的掌握。可以参考 3.2.2 节的开发程序。
【核心代码】本题的核心代码如下所示。

```xml
<EditText
    android:id="@+id/EditText1"
    android:layout_width="wrap_content"
    android:layout_height="wrap_content"
    android:layout_alignParentLeft="true"
    android:layout_alignParentTop="true"
    android:inputType="textPassword"
    android:hint="请输入密码">
</EditText>
<EditText
    android:id="@+id/EditText2"
    android:layout_width="wrap_content"
    android:layout_height="wrap_content"
    android:layout_alignParentLeft="true"
    android:layout_below="@+id/EditText1"
    android:layout_marginTop="26dp"
    android:maxLength="11"
    android:inputType="phone"
    android:lines="1" />
```

2. 新建项目 Button，在布局中添加两个按钮。Button1 在代码中绑定监听，修改标题内容为"Button1 注册成功"；Button2 在布局中通过 onclick 属性绑定监听，修改标题内容为"Button2 注册成功"。程序运行效果如图 3.20 所示。

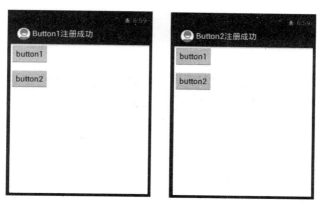

图 3.20　Button 注册监听

【分析】本题目主要考查读者对 Button 两种注册监听方式的掌握。可以参考 3.3.1 节的内容。
【核心代码】本题的核心代码如下所示。
布局文件代码：

```xml
<Button
    android:id="@+id/button1"
    android:layout_width="wrap_content"
    android:layout_height="wrap_content"
    android:text="button1"
    />
<Button
    android:id="@+id/button2"
```

```
    android:layout_width="wrap_content"
    android:layout_height="wrap_content"
    android:layout_alignParentLeft="true"
    android:layout_marginTop="60dp"
    android:onClick="click"
    android:text="button2" />
```

逻辑代码：

```
Button button1,button2;
@Override
public void onCreate(Bundle savedInstanceState) {
    super.onCreate(savedInstanceState);
    setContentView(R.layout.activity_main);
    button1 = (Button)findViewById(R.id.button1);
    button2 = (Button)findViewById(R.id.button2);
    button1 .setOnClickListener(new OnClickListener() {
        public void onClick(View v) {
            // TODO Auto-generated method stub
            setTitle("Button1 注册成功");
        }
    });
}
public void click(View v) {
    setTitle("Button2 注册成功");
}
```

3. 新建项目 RadioButton。在布局中添加一对<RadioGroup></RadioGroup>，设置为水平显示方式。在 RadioGroup 中添加两个 RadioButton，分别代表铃声和振动，供用户选择。为 RadioGroup 绑定监听，使用 TextView 显示用户选中的选项。程序运行效果如图 3.21 所示。

图 3.21　RadioButton

【分析】本题目主要考查读者对 RadioButton、RadioButton 以及其监听事件的掌握。可参考 3.3.4 节的内容。

【核心代码】本题的核心代码如下所示。

布局代码：

```
<RadioGroup
    android:id="@+id/rg1"
    android:layout_width="wrap_content"
    android:layout_height="wrap_content"
```

```xml
        android:orientation="horizontal" >
    <RadioButton
        android:id="@+id/rb1"
        android:layout_width="wrap_content"
        android:layout_height="wrap_content"
        android:textSize="30sp"
        android:text="铃声" />
    <RadioButton
        android:id="@+id/rb2"
        android:layout_width="wrap_content"
        android:layout_height="wrap_content"
        android:textSize="30sp"
        android:text="振动" />
</RadioGroup>
```

逻辑代码:

```java
        radioGroup.setOnCheckedChangeListener(new RadioGroup.OnCheckedChangeListener() {
            public void onCheckedChanged(RadioGroup group, int checkedId) {
                // TODO Auto-generated method stub
                if (checkedId == R.id.rb1) {
                    textView.setText("您选择的是: " + radioButton1.getText());
                }else {
                    textView.setText("您选择的是: " + radioButton2.getText());
                }
            }
        });
```

4. 新建项目 CheckBox。在布局中添加 3 个 CheckBox，分别代表铃声、振动和静音，供用户选择。在 CheckBoxActivity 中，为各个 CheckBox 绑定监听，使用 TextView 显示用户选中的选项，如图 3.22 所示。

图 3.22 CheckBox

布局代码:

```xml
<CheckBox
    android:id="@+id/checkBox1"
    android:layout_width="wrap_content"
    android:layout_height="wrap_content"
    android:textSize="25sp"
    android:text="铃声 />
```

```xml
<CheckBox
    android:id="@+id/checkBox2"
    android:layout_width="wrap_content"
    android:layout_height="wrap_content"
    android:textSize="25sp"
    android:text="振动" />
<CheckBox
    android:id="@+id/checkBox3"
    android:layout_width="wrap_content"
    android:layout_height="wrap_content"
    android:textSize="25sp"
    android:text="静音" />
```

逻辑代码：

```java
checkBox1.setOnCheckedChangeListener(new OnCheckedChangeListener()
{
    public void onCheckedChanged(CompoundButton buttonView,
    boolean isChecked) {
        if (isChecked == true) {
            textView.append(checkBox1 .getText() + ",");
        }
    }
});
checkBox2.setOnCheckedChangeListener(new OnCheckedChangeListener()
{
    public void onCheckedChanged(CompoundButton buttonView,
    boolean isChecked) {
        if (isChecked == true) {
            textView.append(checkBox2 .getText() + ",");
        }
    }
});
checkBox3.setOnCheckedChangeListener(new OnCheckedChangeListener()
{
    public void onCheckedChanged(CompoundButton buttonView,
    boolean isChecked){
        if (isChecked == true) {
            textView.append(checkBox3 .getText() + ",");
        }
    }
});
```

第 4 章 Android 高级控件

在第 3 章中我们学习了 Android 系统中界面的基本控件，如按钮、文本框等。本章我们将继续学习 Android 界面中的一些高级控件，如进度条（ProgressBar）、拖动条（SeekBar）、网格视图（GridView）等。通过这些高级控件，不仅可以使程序界面呈现多样化，也给用户提供了更加友好的用户体验。本章将详细介绍这些控件。

4.1 进度条 ProgressBar

程序处理某些大的数据，在加载这些数据时会一致停在某一界面，这时最好使用进度条。进度条的用途很多，如在登录时比较慢，可以通过进度条进行提示进度；同时也可以对窗口设置进度条。在 Paletee 面板中，提供了 4 种样式的 ProgressBar，分别是 Large、Normal、Small 和 Horizontal，如图 4.1 所示。

ProgressBar 的相关属性如表 4-1 所列。表中 style 属性无需手动设置，从 Paletee 面板中拖动使用 ProgressBar 即会自动生成。Normal 样式的进度条没有 style 属性，其他样式的进度条的 style 属性值如下所述。

- Large：style="?android:attr/progressBarStyleLarge"
- Small：style="?android:attr/progressBarStyleSmall"
- Horizontal：style="?android:attr/progressBarStyleHorizontal"

图 4.1 各式 ProgressBar

表 4-1 ProgressBar 相关属性表

属性名称	属性说明	属性名称	属性说明
style	设置进度条的样式	progress	第一进度值
max	进度条的最大进度值	secordaryProgress	次要进度值

ProgressBar 的语法格式如下：

```
<ProgressBar
    android:id=" "
    <!-- ProgressBar 的宽度和高度-->
    android:layout_width=" "
    android:layout_height=" "
    <!--设置进度条样式-->
    style=" "
    <!--最大进度值-->
    android:max=" "
    <!--第一进度值-->
```

```
android:progress=" "
android:secondaryProgress=" "/>
```

【示例 4-1】 ProgressBar 的使用。新建项目 ProgressBar，在布局文件中添加 4 个不同样式的 ProgressBar 控件。设置水平样式进度条最大值为 100，第一进度值为 75，第二进度值为 50。运行程序，效果如图 4.2 所示。

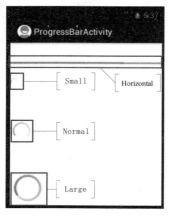

控件	属性	值
ProgressBar (Horizontal)	style	?android:attr/progressBarStyleHorizontal
	max	100
	progress	75
	secordaryProgress	50
ProgressBar (Small)	style	?android:attr/progressBarStyleSmall
ProgressBar (Large)	style	?android:attr/progressBarStyleLarge

图 4.2 ProgressBar

布局代码如下：

```
<ProgressBar
    android:id="@+id/progressBar1"
    style="?android:attr/progressBarStyleHorizontal"
    android:layout_width="wrap_content"
    android:layout_height="wrap_content"
    android:layout_alignParentLeft="true"
    android:layout_alignParentRight="true"
    android:layout_alignParentTop="true"
    android:layout_marginTop="36dp"
    android:max="100"
    android:progress="75"
    android:secondaryProgress="50" />

<ProgressBar
    android:id="@+id/progressBar2"
    style="?android:attr/progressBarStyleSmall"
    android:layout_width="wrap_content"
    android:layout_height="wrap_content"
    android:layout_alignParentLeft="true"
    android:layout_below="@+id/progressBar1"
    android:layout_marginTop="24dp" />

<ProgressBar
    android:id="@+id/progressBar3"
    android:layout_width="wrap_content"
    android:layout_height="wrap_content"
    android:layout_alignParentLeft="true"
    android:layout_below="@+id/progressBar2"
    android:layout_marginTop="76dp" />

<ProgressBar
    android:id="@+id/progressBar4"
    style="?android:attr/progressBarStyleLarge"
    android:layout_width="wrap_content"
```

```
        android:layout_height="wrap_content"
        android:layout_alignParentLeft="true"
        android:layout_below="@+id/progressBar3"
        android:layout_marginTop="62dp" />
```

4.2 拖动条 SeekBar

拖动条（SeekBar）就是添加了滑块的进度条，用户可以通过拖动滑块来调节当前进度。例如，我们可以拖动滑块，调节电影的播放进度，或者调节音量的大小。为了让程序能响应拖动条滑块位置的改变，程序可以考虑为它绑定一个 OnSeekBarChangeListener 监听器。SeekBar 的相关属性如表 4-2 所示。

表 4-2 SeekBar 相关属性表

属性名称	属性说明
android:thumb	设置滑块的样式，值为图片引用
android:max	设置拖动条进度最大值
android:progress	设置拖动条当前进度值

SeekBar 的语法格式如下：

```
<SeekBar
        <!--拖动条 SeekBar ID -->
        android:id=" "
        <!--滑块样式-->
        android:thumb=" "
        <!--拖动条最大值-->
        android:max=" "
        <!--拖动条当前进度值-->
        android:progress=" "
        />
```

【示例 4-2】 SeekBar 的使用。新建项目 SeekBar，在布局文件中添加一个 SeekBar 控件和一个 TextView。设置其最大值为 100，当前值为 25。向右拖动滑块时值增大，向左拖动时值减小，并使用 TextView 显示当前值。运行程序，效果如图 4.3 所示。

布局代码如下：

```
    <SeekBar
        android:id="@+id/seekBar1"
        android:layout_width="match_parent"
        android:layout_height="wrap_content"
        android:layout_alignParentLeft="true"
        android:layout_alignParentTop="true"
        android:layout_marginTop="93dp"
        android:thumb="@android:drawable/btn_star_big_on"
        android:max="100"
        android:progress="25"
        />
    <TextView
        android:id="@+id/textView1"
        android:layout_width="wrap_content"
        android:layout_height="wrap_content"
        android:layout_alignParentLeft="true"
        android:layout_below="@+id/seekBar1"
```

```
android:layout_marginLeft="40dp"
android:text="当前进度值为：25"/>
```

关键代码如下：

```
seekBar.setOnSeekBarChangeListener(new OnSeekBarChangeListener() {
    //为 SeekBar 注册监听
    public void onStopTrackingTouch(SeekBar seekBar) {
        // TODO Auto-generated method stub
    }

    public void onStartTrackingTouch(SeekBar seekBar) {
        // TODO Auto-generated method stub
    }

    public void onProgressChanged(SeekBar seekBar, int progress,
        boolean fromUser) {
        // TODO Auto-generated method stub
        // SeekBar 变化时 TextView 动态显示当前值
        textView.setText("当前进度值: " + progress);
    }
});
```

控件	属性	值
SeekBar	id	@+id/seekBar1
	thumb	@android:drawable/btn_star_big_on
	max	100
	progress	25
TextView	id	@+id/textView1
	text	当前进度值：25

图 4.3　SeekBar

向左拖动滑块时值减小，向右拖动时值增大，拖动到最右端时为最大值 100，如图 4.4 所示。

图 4.4　拖动修改 SeekBar 的值

4.3 自动完成文本控件

在 Android 中提供了两种智能输入框——AutoCompleteTextView 和 MultiAutoCompleteTextView。它们的功能大致相同，类似于百度或者 Google 在搜索栏输入信息的时候，弹出与输入信息接近的提示信息，然后用户选择需要的信息，自动完成文本输入。

4.3.1 使用 AutoCompleteTextView

AutoCompleteTextView（自动完成文本控件）是一个可编辑的文本视图，能够实现动态匹配输入的内容。当用户输入信息后弹出提示信息，提示列表显示在一个下拉菜单中，用户可以从中选择一项自己需要的内容，以完成输入。提示列表是从一个数据适配器获取的数据。工作机制如图 4.5 所示。

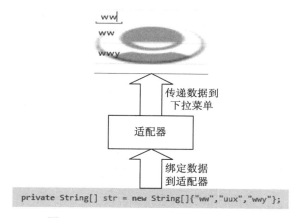

图 4.5 AutoCompleteTextView 工作机制

AutoCompleteTextView 常用属性及对应方法如表 4-3 所示。

表 4-3 AutoCompleteTextView 的属性及对应方法

属性名称	对应方法	属性说明
android:completionThreshold	setThreshold(int)	定义需要用户输入的字符数
android:dropDownHeight	setDropDownHeight(int)	设置下拉菜单高度
android:dropDownWidth	setDropDownWidth(int)	设置下拉菜单宽度
android:popupBackground	setDropDownBackgroundResource(int)	设置下拉菜单背景

AutoCompleteTextView 属性示意图如图 4.6 所示。
AutoCompleteTextView 语法格式如下：

```
<AutoCompleteTextView
    <!-- AutoCompleteTextView 的 ID -->
    android:id=" "
    <!-- AutoCompleteTextView 的宽度和高度-->
    android:layout_width=" "
    android:layout_height=" "
    <!--用户需要输入的字符数-->
```

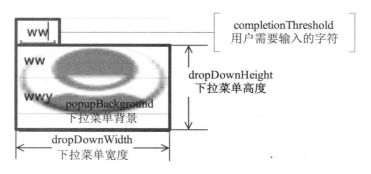

图 4.6 AutoCompleteTextView 属性示意图

```
android:completionThreshold=" "
<!--下拉菜单的高度和宽度-->
android:dropDownHeight=" "
android:dropDownWidth=" "
<!--下拉菜单背景图片-->
android:popupBackground=" "/>
```

【示例 4-3】 AutoCompleteTextView 的使用。新建项目 AutoCompleteTextView，在布局文件中添加一个 AutoCompleteTextView 控件，设置相应属性。当用户连续输入两个"w"之后，出现下拉菜单提示信息，用户可以根据需要进行选择，完成文本的自动输入。

布局代码如下：

```xml
<AutoCompleteTextView
    android:id="@+id/autoCompleteTextView1"
    android:layout_width="wrap_content"
    android:layout_height="wrap_content"
    android:completionThreshold="2"
    android:dropDownHeight="100dp"
    android:dropDownWidth="200dp"
    android:popupBackground="@drawable/ic_launcher"/>
```

逻辑代码如下：

```java
public class AutoCompleteTextViewActivity extends Activity {
    private AutoCompleteTextView autoCompleteTextView;
    // String 数组存放下拉菜单显示的内容
    private String[] str = new String[]{"ww","uux","wwy"};
    @Override
    public void onCreate(Bundle savedInstanceState) {
        super.onCreate(savedInstanceState);
        setContentView(R.layout.activity_auto_complete_text_view);
        autoCompleteTextView =
(AutoCompleteTextView)findViewById(R.id.autoCompleteTextView1);
        //创建适配器向下拉菜单提供数据
        ArrayAdapter<String> adapter = new ArrayAdapter<String>(this,
            android.R.layout.simple_list_item_1,str);
        //为 AutoCompleteTextView 绑定适配器
        autoCompleteTextView.setAdapter(adapter);
    }
}
```

运行程序，效果如图 4.7 所示。

控件	属性	值
AutoComplete TextView	id	@+id/autoCompleteTextView
	completionThreshold	2
	dropDownHeight	100dp
	dropDownWidth	200dp
	popupBackground	@drawable/ic_launcher

图 4.7　AutoCompleteTextView

4.3.2　使用 MultiAutoCompleteTextView

MultiAutoCompleteTextView（多文本自动完成输入控件）也是一个可编辑的文本视图，能够对用户输入的文本进行有效地扩充提示，不需要用户输入完整的内容。用户必须提供一个 MultiAutoCompleteTextView.Tokenizer 用来区分不同的子串。与 AutoCompleteTextView 不同的是，MultiAutoCompleteTextView 可以在输入框一直增加选择值。工作机制如图 4.8 所示。

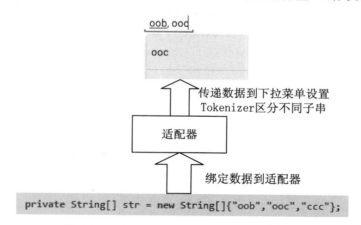

图 4.8　MultiAutoCompleteTextView 工作机制

用户可以在 XML 文件中使用属性进行自动完成文本框的设置，也可以在 Java 代码中通过方法进行设置，下面是常用属性与方法的对照表，如表 4-4 所示。如图 4.9 所示是 MultiAutoCompleteTextView 属性示意图。

表 4-4　MultiAutoCompleteTextView 的属性及对应方法

属性名称	对应方法	属性说明
android:completionThreshold	setThreshold(int)	定义需要用户输入的字符数
android:dropDownHeight	setDropDownHeight(int)	设置下拉菜单高度
android:dropDownWidth	setDropDownWidth(int)	设置下拉菜单宽度
android:popupBackground	setDropDownBackgroundResource(int)	设置下拉菜单背景

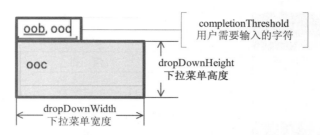

图 4.9　MultiAutoCompleteTextView 属性示意图

MultiAutoCompleteTextView 语法格式如下：

```
<MultiAutoCompleteTextView
    <!-- MultiAutoCompleteTextView 的 ID -->
    android:id=" "
    <!-- MultiAutoCompleteTextView 的宽度和高度-->
    android:layout_width=" "
    android:layout_height=" "
    <!--指定输入几个字符后出现提示信息-->
    android:completionThreshold=" "
    <!--下拉菜单的高度和宽度-->
    android:dropDownHeight=" "
    android:dropDownWidth=" "
    <!--下拉菜单背景图片-->
    android:popupBackground=" " />
```

【示例 4-4】　MultiAutoCompleteTextView 的使用。新建项目 MultiAutoComplete-TextView，在布局文件中添加一个 MultiAutoCompleteTextView 控件，设置相应属性。当用户连续输入两个 "o" 之后，出现下拉菜单提示信息，用户可以根据需要进行选择，完成文本的自动输入。

布局代码如下：

```
<MultiAutoCompleteTextView
    android:id="@+id/multiAutoCompleteTextView1"
    android:layout_width="wrap_content"
    android:layout_height="wrap_content"
    android:completionThreshold="2"
    android:dropDownHeight="100dp"
    android:dropDownWidth="200dp"/>
```

逻辑代码如下：

```
public class MultiAutoCompleteActivity extends Activity {
    private MultiAutoCompleteTextView mAutoCompleteTextView;
    // String数组存放下拉菜单显示的内容
    private String[] str = new String[]{"oob","ooc","ccc"};
    @Override
    public void onCreate(Bundle savedInstanceState) {
        super.onCreate(savedInstanceState);
        setContentView(R.layout.activity_multi_auto_complete);
        mAutoCompleteTextView = (MultiAutoCompleteTextView)
            findViewById(R.id.multiAutoCompleteTextView1);
        //创建适配器,向下拉菜单提供数据
        ArrayAdapter<String> adapter = new ArrayAdapter<String>(this,
            android.R.layout.simple_dropdown_item_1line, str);
```

```
            //为AutoCompleteTextView绑定适配器
            mAutoCompleteTextView.setAdapter(adapter);
            //设置Tokenizer，用来区分不同的子串
            mAutoCompleteTextView.setTokenizer(new
   MultiAutoCompleteTextView.CommaTokenizer());
        }
}
```

运行程序，效果如图 4.10 所示。

控件	属 性	值
MultiAuto-CompleteTextView	id	@+id/multiAutoCompleteTextView1
	dropDownHeight	100dp
	dropDownWidth	200dp

图 4.10　MultiAutoCompleteTextView

4.4　评分条 RatingBar

评分条（RatingBar）是基于 SeekBar 和 ProgressBar 的扩展，用星星来显示等级评定。默认显示 5 颗星，用户可以通过触屏单击或者左右移动轨迹球来进行星星等级评定。RatingBar 相关属性如表 4-5 所列。

表 4-5　RatingBar 相关属性表

属 性 名 称	属 性 说 明
style	RatingBar 样式
android:isIndicator	RatingBar 是否是一个指示器（值为 true 时，用户无法进行更改）
android:numStars	显示的星星数量，必须是一个整型值
android:rating	默认的评分，必须是浮点类型
android:stepSize	评分的步长，即一次增加或者减少的星星数目是这个数字的整数倍，必须是浮点类型

RatingBar 有 3 种风格，即 RatingBarStyle（默认风格）、RatingBarStyleSmall（小风格）和 RatingBarStyleIdicator（大风格）。其中，默认风格为 RatingBarStyle，是我们通常使用的、可以交互的风格，而后面两种风格不能进行进行交互，只能作为指示牌。

设置 RatingBar 样式的方法是在布局文件中设置 RatingBar 控件的 style 属性：

```
style="?android:attr/ratingBarStyle"
```

```
style="?android:attr/ratingBarStyleSmall"
style="?android:attr/ratingBarStyleIndicator"
```

效果如图 4.11 所示。

图 4.11　各式 RatingBar

RatingBar 语法格式如下：

```
<RatingBar
<!-- MultiAutoCompleteTextView 的 ID -->
    android:id=" "
    <!--指定输入几个字符后出现提示信息-->
    style=" "
    <!-- MultiAutoCompleteTextView 的宽度和高度-->
    android:layout_width=" "
    android:layout_height=" "
    <!--指定输入几个字符后出现提示信息-->
    android:numStars=" "
    <!--指定输入几个字符后出现提示信息-->
    android:isIndicator=" "
    <!--指定输入几个字符后出现提示信息-->
    android:rating=" "

    android:stepSize=" " />
```

【示例 4-5】RatingBar 的使用。新建项目 RatingBar，在布局文件中添加一个 RatingBar 控件和一个 TextView。设置当前评分是 4 颗星，评分步长是 0.5。单击右侧星星评分增大，单击左侧星星评分减小，使用 TextView 显示当前分值。运行程序，效果如图 4.12 所示。

布局代码如下：

```
<RatingBar
    android:id="@+id/ratingBar1"
    style="?android:attr/ratingBarStyle"
    android:layout_width="wrap_content"
    android:layout_height="wrap_content"
    android:layout_alignParentLeft="true"
    android:layout_alignParentTop="true"
    android:layout_marginTop="14dp"
    android:numStars="5"
```

```xml
        android:rating="4.0"
        android:stepSize="0.5"/>

<TextView
    android:id="@+id/textView1"
    android:layout_width="wrap_content"
    android:layout_height="wrap_content"
    android:layout_below="@+id/ratingBar1"
    android:layout_marginTop="20dp"
    android:text="受欢迎度：4.0颗星"/>
```

关键代码如下：

```java
ratingBar.setOnRatingBarChangeListener(new OnRatingBarChangeListener(){
    //为 RatingBar 注册监听
    public void onRatingChanged(RatingBar ratingBar, float rating,
            boolean fromUser) {
        // TODO Auto-generated method stub
        // RatingBar 变化时，TextView 动态显示当前值
        textView.setText("受欢迎度为：" + rating + "颗星");
    }
});
```

控件	属性	值
RatingBar	id	@+id/ratingBar1
	style	?android:attr/ratingBarStyle
	numStars	5
	rating	4.0
	stepSize	0.5
TextView	id	@+id/textView1
	text	受欢迎度：4.0颗星

图 4.12　RatingBar

单击右侧星星评分增大，单击左侧星星评分减小，如图 4.13 所示。

图 4.13　单击修改 RatingBar 的值

4.5 下拉列表 Spinner

下拉列表（Spinner）每次只显示用户选中的元素，当用户再次单击时，会弹出选择列表供用户选择，而选择列表中的元素同样是来自适配器。Spinner 相关属性及对应方法如表 4-6 所示。

表 4-6　Spinner 的常用属性及对应方法表

属 性 名 称	对 应 方 法	属 性 说 明
android:spinnerMode		设置 Spinner 样式，有 dropdown 和 dialog 两种
android:dropDownVerticalOffset	setDropDownVerticalOffset(int)	设置 Spinner 下拉菜单的水平偏移
android:dropDownHorizontalOffset	setDropDownHorizontalOffset(int)	设置 Spinner 下拉菜单的垂直偏移
android:dropDownWidth	setDropDownWidth(int)	设置 Spinner 下拉菜单的宽度
android:popupBackground	setPopupBackgroundResource()	设置下拉菜单的背景

Spinner 语法格式如下：

```xml
<Spinner
    <!-- MultiAutoCompleteTextView 的 ID -->
    android:id=" "
    <!--指定输入几个字符后出现提示信息-->
    android:layout_width=" "
    android:layout_height=" "
    <!-- MultiAutoCompleteTextView 的宽度和高度-->
    android:spinnerMode=" "
    <!--指定输入几个字符后出现提示信息-->
    android:dropDownWidth=" "
    <!--指定输入几个字符后出现提示信息-->
    android:dropDownVerticalOffset=" "
    <!--指定输入几个字符后出现提示信息-->
    android:dropDownHorizontalOffset=" "

    android:popupBackground=" "/>
```

【示例 4-6】 Spinner 的使用。新建项目 Spinner，在布局文件中添加一个 Spinner 控件。使用 Spinner 实现选择 character 的功能。运行程序，效果如图 4.14 所示。

布局代码如下：

```xml
<Spinner
    android:id="@+id/spinner1"
    android:layout_width="fill_parent"
    android:layout_height="wrap_content"
    android:spinnerMode="dropdown"
    android:dropDownWidth="wrap_content"
    android:dropDownVerticalOffset="20dp"
    android:dropDownHorizontalOffset="10dp"
    android:popupBackground="@drawable/ic_launcher"/>
```

逻辑代码如下：

```java
public class SpinnerActivity extends Activity {
    private Spinner spinner;
    //下拉列表数据
    private String character[]={
            "A",
            "B",
            "C",
            "D",
            "E"};
    //声明List,作适配器参数
    private List<String> list = new ArrayList<String>();
    @Override
    public void onCreate(Bundle savedInstanceState) {
        super.onCreate(savedInstanceState);
        setContentView(R.layout.activity_spinner);
        spinner=(Spinner)findViewById(R.id.spinner1);
        //添加数组元素为List值
        for (int i = 0; i < character.length; i++) {
            list.add(character[i]);
        }
        //创建适配器向ListView提供数据
        ArrayAdapter<String> adapter = new ArrayAdapter<String>
        (this, android.R.layout.simple_spinner_dropdown_item,list);
        //为Spinner绑定适配器
        spinner.setAdapter(adapter);
    }
}
```

图 4.14　Spinner

控件	属性	值
Spinner	id	@+id/spinner1
	spinnerMode	dropdown
	dropDownVerticalOffset	20dp
	dropDownHorizontalOffset	10dp
	android:dropDownWidth	warp-content
	android:popupBackground	@drawable/ic_launcher

4.6　选项卡 TabHost

选项卡（TabHost）控件可以在一个屏幕间进行不同版面的切换。单击每个选项卡，打开其对应的内容界面。TabHost 是整个 Tab 的容器，包括 TabWidget 和 FrameLayout 两部分。TabWidget 就是每个 Tab 的标签，FrameLayout 则是 Tab 的内容。

TabHost 语法格式如下：

```xml
<TabHost
    xmlns:android="http://schemas.android.com/apk/res/android"
    <!-- TabHost 的宽度和高度-->
    android:layout_width=" "
    android:layout_height=" "
    <!-- ListView 分割线-->
    android:id="@android:id/tabhost"
>
<!-- Tab 标签固定 ID -->
<TabWidget
    android:id="@android:id/tabs"
    android:layout_width=" "
    android:layout_height=" " >
</TabWidget>
<!-- Tab 内容-->
<FrameLayout
    …… >
</FrameLayout>
</TabHost>
```

【示例 4-7】 TabHost 的使用。新建项目 TabHost，将默认布局 Relativelayout 修改为 Tabhost。然后并列添加 TabWidget 标签和 FrameLayout 布局，TabWidget 用来显示 Tab 标签，FrameLayout 用来显示 Tab 内容。

布局代码如下：

```xml
<?xml version="1.0" encoding="utf-8"?>
<TabHost xmlns:android="http://schemas.android.com/apk/res/android"
    android:layout_width="fill_parent"
    android:layout_height="fill_parent"
    android:id="@android:id/tabhost"
>

<LinearLayout android:id="@+id/tab11"
    android:orientation="vertical"
    android:layout_width="fill_parent"
    android:layout_height="fill_parent"
    >

<TabWidget
    android:id="@android:id/tabs"
    android:layout_width="fill_parent"
    android:layout_height="wrap_content" >
</TabWidget>
<FrameLayout
    android:id="@android:id/tabcontent"
    android:layout_width="fill_parent"
    android:layout_height="fill_parent"
    android:layout_weight="1" >

    <TextView
        android:id="@+id/tv11"
        android:layout_width="wrap_content"
        android:layout_height="wrap_content"
        android:text="TAB1"
```

```xml
            android:textSize="11pt" />

        <TextView
            android:id="@+id/tv22"
            android:layout_width="wrap_content"
            android:layout_height="wrap_content"
            android:text="TAB2 "
            android:textSize="11pt" />

        <TextView
            android:id="@+id/tv33"
            android:layout_width="wrap_content"
            android:layout_height="wrap_content"
            android:text="TAB3 "
            android:textSize="11pt" />
    </FrameLayout>
</LinearLayout>
</TabHost>
```

逻辑代码如下:

```java
public class TabHostActivity extends Activity {
    @Override
    public void onCreate(Bundle savedInstanceState) {
        super.onCreate(savedInstanceState);
        setContentView(R.layout.activity_tab_host);
        TabHost tabhost = (TabHost)findViewById(android.R.id.tabhost);
        //必须调用该方法,才能设置 Tab 样式
        tabhost.setup();
        //添加标签 tab1
        tabhost.addTab(tabhost.newTabSpec("tab1")
                //设置 tab1 标签图片
                .setIndicator(null,
getResources().getDrawable(R.drawable.png1144))
                //设置 tab1 内容
                .setContent(R.id.tv11));
        //添加标签 tab2
        tabhost.addTab(tabhost.newTabSpec("tab2")
                //设置 tab2 标签图片
                .setIndicator(null,
getResources().getDrawable(R.drawable.png1145))
                //设置 tab2 内容
                .setContent(R.id.tv22));
        //添加标签 tab3
        tabhost.addTab(tabhost.newTabSpec("tab3")
                //设置 tab3 标签图片
                .setIndicator(null,
getResources().getDrawable(R.drawable.png1148))
                //设置 tab3 内容
                .setContent(R.id.tv33));
        //设置当前显示第一个 tab
        tabhost.setCurrentTab(0);
    }
}
```

运行程序,效果如图 4.15 所示,单击 Tab 标签切换不同版面。

图 4.15　TabHost

4.7　图片切换控件 ImageSwitcher

ImageSwitcher 是 Android 中控制图片展示效果的控件。类似于 Window 图片和传真查看器，在"下一张"和"上一张"之间切换显示图片。

ImageSwitcher 类必须设置一个 ViewFactory，用来将显示的图片和父窗口区分开来，因此需要实现 ViewSwitcher.ViewFactory 接口。通过 makeView()方法来显示图片，会返回一个 ImageView 对象，而方法 setImageResource 用来指定图片资源。ImageSwitcher 相关属性如表 4-7 所示。

表 4-7　ImageSwitchery 常用属性

属 性 名 称	属 性 说 明
android:animateFirstView	定义 ViewAnimation 首次显示时是否对当前视图应用动画
android:inAnimation	标识显示视图时使用的动画
android:outAnimation	标识隐藏视图时使用的动画

ImageSwitcher 语法格式如下：

```xml
<ImageSwitcher
<!-- ImageSwitcher 的 ID -->
    android:id=" "
    <!-- ImageSwitcher 的宽度和高度-->
    android:layout_width=" "
    android:layout_height=" "
    <!--首次显示时是否对当前视图应用动画-->
    android:animateFirstView=" "
    <!--标识显示视图时使用的动画-->
    android:inAnimation=" "
    <!--标识隐藏视图时使用的动画-->
    android:outAnimation=" ">

</ImageSwitcher>
```

【示例 4-8】 ImageSwitcher 的使用。新建项目 ImageSwitcher，在布局文件中添加一个

ImageSwitcher 控件显示图片，添加两个 Button，分别为"上一张"和"下一张"。单击按钮，切换显示图片。

布局代码如下：

```xml
<ImageSwitcher
    android:id="@+id/imageSwitcher1"
    android:layout_width="200px"
    android:layout_height="200px"
    android:layout_alignParentTop="true"
    android:layout_centerHorizontal="true"
    android:animateFirstView="true"
    android:inAnimation="@android:anim/fade_in"
    android:outAnimation="@android:anim/fade_out">
</ImageSwitcher>

<Button
    android:id="@+id/button1"
    android:layout_width="wrap_content"
    android:layout_height="wrap_content"
    android:layout_alignRight="@+id/imageSwitcher1"
    android:layout_below="@+id/imageSwitcher1"
    android:text="下一张" />

<Button
    android:id="@+id/button2"
    android:layout_width="wrap_content"
    android:layout_height="wrap_content"
    android:layout_alignBaseline="@+id/button1"
    android:layout_alignBottom="@+id/button1"
    android:layout_alignLeft="@+id/imageSwitcher1"
    android:text="上一张" />
```

逻辑代码如下：

```java
public class ImageSwitcherActivity extends Activity {
    private Button buttonPre,buttonNext;
    private ImageSwitcher imageSwitcher;
    //图片 id 数组
    private int[] imageIds = {
            R.drawable.png1132,
            R.drawable.png1139,
            R.drawable.png1144,
            R.drawable.png1145,
            R.drawable. png1148 };
    //图片索引
    private int index=0;
    @Override
    public void onCreate(Bundle savedInstanceState) {
        super.onCreate(savedInstanceState);
        setContentView(R.layout.activity_image_switcher);
        imageSwitcher = (ImageSwitcher)findViewById(R.id.imageSwitcher1);
        //实现 ViewFactory 接口
        imageSwitcher.setFactory(new ViewFactory() {
            //调用 makeView()方法设置图片显示效果,返回 ImageView 对象
            public View makeView() {
                ImageView imageView = new ImageView(ImageSwitcherActivity.this);
                imageView.setBackgroundColor(0xff0000);
                imageView.setScaleType(ImageView.ScaleType.FIT_CENTER);
```

```java
                imageView.setLayoutParams(new 
ImageSwitcher.LayoutParams(200,200));
                return imageView;
            }
        });
        //根据id设置背景图片
        imageSwitcher.setBackgroundResource(imageIds[index]);
        //"下一张"按钮监听
        buttonNext = (Button)findViewById(R.id.button1);
        buttonNext.setOnClickListener(new OnClickListener() {
            public void onClick(View v) {
                // TODO Auto-generated method stub
                if(index>=0&&index<imageIds.length-1)
                {
                //数组下标加1,显示下一张图片
                 index++;
                 imageSwitcher.setBackgroundResource(imageIds[index]);
                }else
                {
                 index=imageIds.length-1;
                }
            }
        });
        //"上一张"按钮监听
        buttonPre = (Button)findViewById(R.id.button2);
        buttonPre.setOnClickListener(new OnClickListener() {
            public void onClick(View v) {
                // TODO Auto-generated method stub
                if(index>0&&index<imageIds.length)
                {
                //数组下标减1,显示上一张图片
                 index--;
                 imageSwitcher.setBackgroundResource(imageIds[index]);
                }else
                {
                 index=imageIds.length-1;
                }
            }
        });
    }
}
```

运行程序,效果如图4.16所示,单击按钮切换显示图片。

图4.16　ImageSwitcher

4.8 列表视图 ListView

列表视图（ListView）是将数据显示在一个垂直且可滚动的列表中的一种控件。数据可以引用 value 目录下的 arrays.xml 数组元素，也可以引用代码中自定义的数据元素，由与 ListView 绑定的 ListAdapter 传递。每一行数据为一条 item。ListView 的相关属性如表 4-8 所示。

表 4-8 ListView 相关属性表

属 性 名 称	属 性 说 明
android:divider	每条 item 之间的分割线，参数值可引用一张 drawable 图片，也可以是 color
android:dividerHeight	分割线的高度
android:entries	引用一个将使用在此 ListView 里的数组，该数组定义在 value 目录下的 arrays.xml 文件中
android:footerDividersEnabled	设为 flase 时，此 ListView 将不会在页脚视图前画分隔符，默认值为 true
android:headerDividersEnabled	设为 flase 时，此 ListView 将不会在页眉视图后画分隔符，默认值为 true

ListView 语法格式如下：

```xml
<ListView
  <!-- ListView 的 ID -->
  android:id=" "
  <!-- ListView 的高度和宽度-->
  android:layout_height=" "
  android:layout_width=" "
  <!-- ListView 分割线-->
  android:divider=" "
  <!--分割线高度-->
  android:dividerHeight=" "
  <!-- ListView 引用数组-->
  android:entries=" "
  <!--是否在页脚试图前画分隔符-->
  android:headerDividersEnabled=" "

  android:footerDividersEnabled=" "/>
```

【示例 4-9】 使用 arrays.xml 文件中数组元素的 ListView。新建项目 ListView，在 value 目录下新建 arrays.xml 文件，添加 string-array 数组供 ListView 显示。在布局文件中添加一个 ListView 控件，设置其引用 string-array，并设置分割线以及高度。运行程序，效果如图 4.17 所示。

arrays.xml 文件代码：

```xml
<resources>
  <string-array name="planets">
    <item>Mercury</item>
    <item>Venus</item>
    <item>Earth</item>
    <item>Mars</item>
    <item>Jupiter</item>
```

```
            <item>Saturn</item>
            <item>Uranus</item>
            <item>Neptune</item>
            <item>Pluto</item>
    </string-array>
</resources>
```

控件	属　性	值
ListView	id	@+id/listView1
	dividerHeight	#87CEFF
	entries	@array/planets
	footerDividersEnabled	false
	headerDividersEnabled	false

图 4.17　使用 Arrays.xml 数组元素的 ListView

【示例 4-10】 使用自定义数组元素的 ListView。模拟实现一个电话本，每一条 item 显示一张图片、一个用户名和对应的 QQ 号码。

ListView 布局文件：

```
<ListView
    android:id="@+id/listView1"
    android:layout_height="fill_parent"
    android:layout_width="fill_parent"
    android:divider="#87CEFF"
    android:dividerHeight="2dp"
/>
```

item 布局文件：

```
<ScrollView xmlns:android="http://schemas.android.com/apk/res/android"
    xmlns:tools="http://schemas.android.com/tools"
    android:id="@+id/scollView1"
    android:layout_width="match_parent"
    android:layout_height="match_parent">
<TableLayout xmlns:android="http://schemas.android.com/apk/res/android"
    android:layout_height="fill_parent"
    android:layout_width="fill_parent"
    android:orientation="vertical"
    android:stretchColumns="1"
>

<TableRow >
    <ImageView
        android:layout_width="wrap_content"
        android:id="@+id/imageView1"
        android:layout_height="wrap_content"
        android:src="@drawable/ic_launcher">
    </ImageView>
```

```xml
<TextView
    android:textAppearance="?android:attr/textAppearanceLarge"
    android:id="@+id/name"
    android:layout_height="wrap_content"
    android:layout_width="wrap_content"  >
</TextView>

<TextView
    android:textAppearance="?android:attr/textAppearanceLarge"
    android:id="@+id/qq"
    android:layout_height="wrap_content"
    android:layout_width="wrap_content"  >
</TextView>
    </TableRow>
</TableLayout>
</ScrollView>
```

逻辑代码：

```java
public class ListViewActivity extends Activity {
    //声明 List,作适配器参数
    private List<Map<String, Object>> sList = new ArrayList<Map<String, Object>>();
    //name 数组
    private String name[] = {
            "Ricky",
            "Daisy",
            "Seven",
            "Fever",
            "Mike",
            "Rose",
            "Jack",
            "Tom",
            "Jim"
    };
    //num 数组
    private String num[] = {
            "15535185171",
            "18810448744",
            "13294622375",
            "15536082491",
            "15536823409",
            "13728907653",
            "15136884219",
            "18734628724",
            "13322442390"
    };
    @Override
    public void onCreate(Bundle savedInstanceState) {
        super.onCreate(savedInstanceState);
        setContentView(R.layout.activity_list_view);
        //添加数组元素为 List 值
        for (int i = 0; i < name.length; i++) {
            Map<String, Object> map = new HashMap<String, Object>();
            map.put("userPic", R.drawable.ic_launcher);
            map.put("userName", name[i]);
            map.put("userNum", num[i]);
            sList.add(map);
        }
```

```
    //创建适配器向 ListView 提供数据
    ListAdapter listAdapter = new SimpleAdapter(this,sList,R.layout.
    another_layout,
        new String []{"userPic","userName","userNum"},
        new int[]{R.id.imageView1,R.id.name,R.id.num}
        );
    //为 ListView 绑定适配器
    ((ListView)findViewById(R.id.listView1)).setAdapter(listAdapter);
    }
}
```

运行程序，效果如图 4.18 所示，滚动显示全部内容。

图 4.18　使用自定义数组元素的 ListView

4.9　网格视图 GridView

网格视图（GridView）是 Android 中比较常用的多控件视图。该视图将其他多个控件，以二维格式显示在界面表格中，这些控件都来自于 ListAdapter。GridView 相关属性及对应方法如表 4-9 所示。

表 4-9　网格视图的属性与方法

属性名称	对应方法	属性说明
android:columnWidth	setColumnWidth(int)	设置列的宽度
android:gravity	setGravity(int)	设置对齐方式
android:horizontalSpacing	setHorizontalSpacing(int)	设置各个元素之间的水平距离
android:numColumns	setNumColumns(int)	设置列数
android:verticalSpacing	setVerticalSpacing(int)	设置各个元素之间的竖直距离

GridView 语法格式如下：

```
<GridView
    <!-- GridView 的 ID -->
    android:id=" "
    <!-- GridView 的高度和宽度-->
    android:layout_width=" "
    android:layout_height=" "
```

```xml
<!-- GridView 的列数-->
android:numColumns=" "
<!-- GridView 每列宽度-->
android:columnWidth=" "
<!-- GridView 的对齐方式-->
android:gravity=" "
<!--各控件间垂直距离和水平距离-->
android:verticalSpacing=" "
android:horizontalSpacing=" ">
</GridView>
```

【示例 4-11】 GridView 的使用。新建项目 GridView，添加 GridView 控件；再新建一个布局文件 another_layout.xml，并添加一个 ImageView 控件和一个 TextView 控件，控制每一个网格布局。其中，ImageView 用来显示网格视图中各元素图片，TextView 用来显示对应图片的文本内容。

布局代码如下：

```xml
<GridView
    android:id="@+id/gridView1"
    android:layout_width="match_parent"
    android:layout_height="wrap_content"
    android:numColumns="3"
    android:columnWidth="80dp"
    android:gravity="center_vertical"
    android:verticalSpacing="5dp"
    android:horizontalSpacing="10dip">
</GridView>
```

another_layout.xml 代码：

```xml
<ImageView android:id="@+id/image"
    android:layout_width="80dip"
    android:layout_height="80dip">
</ImageView>

<TextView android:id="@+id/name"
    android:layout_below="@+id/image"
    android:layout_height="wrap_content"
    android:layout_width="wrap_content"
    android:gravity="center">
</TextView>
```

逻辑代码如下：

```java
public class GridViewActivity extends Activity {
    private GridView gridView;
    private List<Map<String, Object>> list;
    //TextView 文本内容
    private String[] name = {"album","clock","dir","msm","music","photo","set","video","web"};
    //图片 id 数组
    private int[] imgId = {
            R.drawable.album,R.drawable.clock,R.drawable.dir,
            R.drawable.msm,R.drawable.music,R.drawable.photo,
            R.drawable.set,R.drawable.video,R.drawable.web,
    };
    @Override
    public void onCreate(Bundle savedInstanceState) {
```

```java
    super.onCreate(savedInstanceState);
    setContentView(R.layout.activity_grid_view);
    gridView = (GridView)findViewById(R.id.gridView1);
    //利用 for 循环,将图片和文本添加到 List 中
    list = new ArrayList<Map<String,Object>>();
    for (int i = 0; i < 9; i++) {
        Map<String, Object> map = new HashMap<String, Object>();
        map.put("image", imgId[i]);
        map.put("name", name[i]);
        list.add(map);
    }
    //声明适配器,为 GridView 传递数据
    SimpleAdapter adapter = new SimpleAdapter(this, list,
        R.layout.another_layout,
        new String[]{"image", "name"},
        new int[]{R.id.image, R.id.name});
    //绑定适配器到 GridView
    gridView.setAdapter(adapter);
    }
}
```

运行程序,效果如图 4.19 所示。

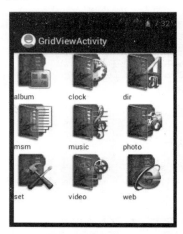

图 4.19　GridView

4.10　小　　结

本章主要介绍了在 Android 平台中的一些高级控件。其中,自动完成文本、进度条、ListView、Spinner、TabHost 和 GridView 是在开发中较为常用的,读者要熟练掌握。本章中多数控件的使用,都涉及了适配器,希望读者细心总结各种适配器的适用控件。

4.11　习　　题

1. 新建项目 MultiAutoCompleteTextView。在布局文件中添加 MultiAutoComplete-TextView 控件,实现多文本自动完成输入功能,如图 4.20 所示。

【分析】本题目主要考查读者对 MultiAutoCompleteTextView 控件的掌握。在使用 MultiAutoCompleteTextView 时,需要适配器提供文本信息。还需要设置 MultiAutoComplete-TextView.Tokenizer 来区分不同的子串。可以参考 4.1.2 节的内容。

【核心代码】本题的核心代码如下所示。

布局代码:

```xml
<MultiAutoCompleteTextView
    android:id="@+id/multiAutoCompleteTextView1"
    android:layout_width="wrap_content"
    android:layout_height="wrap_content"
    android:completionThreshold="2"
    android:dropDownHeight="200dp"
    android:dropDownWidth="200dp"/>
```

逻辑代码:

```java
private String[] str = new String[]{"ece","ecc","ggh"};
ArrayAdapter<String> adapter = new ArrayAdapter<String>(this,
        android.R.layout.simple_dropdown_item_1line, str);
mAutoCompleteTextView.setAdapter(adapter);
mAutoCompleteTextView.setTokenizer(new
MultiAutoCompleteTextView.CommaTokenizer());
```

2. 新建项目,在布局文件中添加一个 SeekBar 显示进度值;添加一个 RatingBar,显示等级。如图 4.21 所示。

图 4.20　MultiAutoCompleteTextView　　　　图 4.21　SeekBar 和 RatingBar

【分析】本题目主要考查读者对 SeekBar 和 RatingBar 控件的掌握,以及这两种控件的监听事件。可以参考 4.3 和 4.4 节的内容。

【核心代码】本题的核心代码如下所示。

布局代码:

```xml
<SeekBar
    android:id="@+id/seekBar1"
    android:layout_width="match_parent"
    android:layout_height="wrap_content"
    android:layout_alignParentLeft="true"
    android:layout_marginTop="21dp"
    android:max="100"
    android:progress="10"/>
```

```xml
<RatingBar
    android:id="@+id/ratingBar1"
    android:layout_width="wrap_content"
    android:layout_height="wrap_content"
    android:layout_alignParentLeft="true"
    android:layout_below="@+id/textView1"
    android:rating="4.5"
    android:layout_marginTop="30dp" />
```

逻辑代码：

```java
seekBar.setOnSeekBarChangeListener(new OnSeekBarChangeListener() {
    public void onStopTrackingTouch(SeekBar seekBar) {
        // TODO Auto-generated method stub
    }
    public void onStartTrackingTouch(SeekBar seekBar) {
        // TODO Auto-generated method stub
    }
    public void onProgressChanged(SeekBar seekBar, int progress,
            boolean fromUser) {
        // TODO Auto-generated method stub
        textView1.setText("当前进度值: " + progress);
    }
});
ratingBar.setOnRatingBarChangeListener(new OnRatingBarChangeListener(){
    public void onRatingChanged(RatingBar ratingBar, float rating,
            boolean fromUser) {
        // TODO Auto-generated method stub
        textView2.setText("受欢迎度为: " + rating + "颗星");
    }
});
```

3. 新建项目 List，在布局文件中添加 ListView 控件，设置每条 item 间的分割线。引用 Value 目录下的 arrays.xml 文件，显示其中的数组元素，运行效果如图 4.22 所示。

【分析】本题目主要考查读者对 arrays.xml 文件中数据的 ListView 的掌握。可以参考 4.5 节的内容。

【核心代码】本题的核心代码如下所示。

布局文件：

```xml
<ListView
    android:id="@+id/listView1"
    android:layout_height="fill_parent"
    android:layout_width="fill_parent"
    android:divider="#87CEFF"
    android:dividerHeight="2dp"
    android:entries="@array/week"
    android:headerDividersEnabled="false"
    android:footerDividersEnabled="false"/>
```

arrays.xml：

```xml
<resources>
    <string-array name="week">
        <item>Monday</item>
        <item>Tuesday</item>
```

```
        <item>Wednesday</item>
        <item>Thursday</item>
        <item>Friday</item>
        <item>Saturday</item>
        <item>Sunday</item>
    </string-array>
</resources>
```

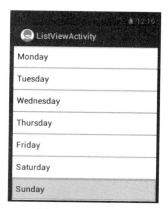

图 4.22　ListView

图 4.23　Spinner

4. 新建项目 Spinner，在布局文件中添加 Spinner 控件。在下拉菜单中选择你的幸运数字，如图 4.23 所示。

【分析】本题目主要考查读者对 Spinner 控件的掌握，注意 Spinner 需要数组适配器传递数据。可以参考 4.6 节的内容。

【核心代码】本题的核心代码如下所示。

布局文件：

```xml
<Spinner
    android:id="@+id/spinner1"
    android:layout_width="fill_parent"
    android:layout_height="wrap_content"
    android:spinnerMode="dropdown"
    android:dropDownWidth="wrap_content"
    android:dropDownVerticalOffset="20dp"
    android:dropDownHorizontalOffset="10dp"
    android:popupBackground="@drawable/ic_launcher"/>
```

逻辑代码：

```java
public class SpinnerActivity extends Activity {
    private Spinner spinner;
    //下拉列表数据
    private String number[]={
                "0",
                "1",
                "2",
                "3",
                "4",
                "5",
                "6",
                "7",
```

```
                "8",
                "9"};
    //声明 List,作适配器参数
    private List<String> list = new ArrayList<String>();
    @Override
    public void onCreate(Bundle savedInstanceState) {
        super.onCreate(savedInstanceState);
        setContentView(R.layout.activity_spinner);
        spinner=(Spinner)findViewById(R.id.spinner1);
        //添加数组元素为 List 值
        for (int i = 0; i < number.length; i++) {
            list.add(number[i]);
        }
        ArrayAdapter<String> adapter = new ArrayAdapter<String>
        (this, android.R.layout.simple_spinner_dropdown_item,list);
        spinner.setAdapter(adapter);
    }
}
```

5. 新建项目 ImageSwitcher，在布局文件中添加 ImageSwitcher 控件，实现图片浏览功能，如图 4.24 所示。

图 4.24　ImageSwitcher

【分析】本题目主要考查读者对 ImageSwitcher 控件的掌握，通过改变存放图片 id 数组的下标浏览图片。可以参考 4.7 节的内容。

【核心代码】本题的核心代码如下所示。

布局文件：

```
<ImageSwitcher
    android:id="@+id/imageSwitcher1"
    android:layout_width="150dp"
    android:layout_height="150dp"
    android:layout_alignParentTop="true"
    android:layout_centerHorizontal="true"
    android:animateFirstView="true"
    android:inAnimation="@android:anim/fade_in"
    android:outAnimation="@android:anim/fade_out" >
</ImageSwitcher>
```

逻辑代码：

```java
//图片id数组
    private int[] imageIds = {
        R.drawable.cope, R.drawable.player,R.drawable.band
        };
//图片索引
    private int index=0;
imageSwitcher = (ImageSwitcher)findViewById(R.id.imageSwitcher1);
    imageSwitcher.setFactory(new ViewFactory() {
            public View makeView() {
                ImageView imageView = new ImageView(ImageSwitcherActivity.this);
                imageView.setBackgroundColor(0xff0000);
                imageView.setScaleType(ImageView.ScaleType.FIT_CENTER);
                imageView.setLayoutParams(new ImageSwitcher.LayoutParams
                (200,200));
                return imageView;
            }
        });
    imageSwitcher.setBackgroundResource(imageIds[index]);
    //"下一张"按钮监听
    buttonNext = (Button)findViewById(R.id.button1);
    buttonNext.setOnClickListener(new OnClickListener() {
            public void onClick(View v) {
                // TODO Auto-generated method stub
                if(index>=0&&index<imageIds.length-1)
                {
                 index++;
                 imageSwitcher.setBackgroundResource(imageIds[index]);
                }else
                {
                 index=imageIds.length-1;
                }
            }
        });
    //"上一张"按钮监听
    buttonPre = (Button)findViewById(R.id.button2);
    buttonPre.setOnClickListener(new OnClickListener() {
            public void onClick(View v) {
                // TODO Auto-generated method stub
                if(index>0&&index<imageIds.length)
                {
                 index--;
                 imageSwitcher.setBackgroundResource(imageIds[index]);
                }else
                {
                 index=imageIds.length-1;
                }
            }
        });
```

6. 新建项目 TabHost。开发一个拥有 3 个 Tab 选项卡的 TabHost，并设置程序启动后，首先显示第一个选项卡内容，如图 4.25 所示。

【分析】本题目主要考查读者对 TabHost 控件的掌握。需要注意的是，该程序的根布局文件是 TabHost，Tab 内容使用 FrameLayout 布局。可以参考 4.8 节的内容。

【核心代码】本题的核心代码如下所示。

布局文件：

图 4.25　TabHost

```xml
<?xml version="1.0" encoding="utf-8"?>
<TabHost xmlns:android="http://schemas.android.com/apk/res/android"
    android:layout_width="fill_parent"
    android:layout_height="fill_parent"
    android:id="@android:id/tabhost">

<LinearLayout android:id="@+id/tab11"
    android:orientation="vertical"
    android:layout_width="fill_parent"
    android:layout_height="fill_parent"
    >

<TabWidget
      android:id="@android:id/tabs"
      android:layout_width="fill_parent"
      android:layout_height="wrap_content" >
    </TabWidget>
    <FrameLayout
      android:id="@android:id/tabcontent"
      android:layout_width="fill_parent"
      android:layout_height="fill_parent"
      android:layout_weight="1" >

      <TextView
        android:id="@+id/tv11"
        android:layout_width="wrap_content"
        android:layout_height="wrap_content"
        android:text="TAB1"
        android:textSize="11pt" />

      <TextView
        android:id="@+id/tv22"
        android:layout_width="wrap_content"
        android:layout_height="wrap_content"
        android:text="TAB2 "
        android:textSize="11pt" />

      <TextView
        android:id="@+id/tv33"
        android:layout_width="wrap_content"
        android:layout_height="wrap_content"
        android:text="TAB3 "
        android:textSize="11pt" />
```

```
    </FrameLayout>
</LinearLayout>
</TabHost>
```

逻辑代码:

```
        TabHost tabhost = (TabHost)findViewById(android.R.id.tabhost);
        tabhost.setup();
        tabhost.addTab(tabhost.newTabSpec("tab1")
              .setIndicator(null , getResources().getDrawable
              (R.drawable.band))
              .setContent(R.id.tv11));
        tabhost.addTab(tabhost.newTabSpec("tab2")
              .setIndicator(null , getResources().getDrawable
              (R.drawable.cope))
              .setContent(R.id.tv22));
        tabhost.addTab(tabhost.newTabSpec("tab3")
              .setIndicator(null , getResources().getDrawable
              (R.drawable. player))
              .setContent(R.id.tv33));
           tabhost.setCurrentTab(0);
```

第 5 章　Android 菜单和对话框

在 Android 中，菜单和对话框的设计对于人机交换是非常人性化的。菜单提供了不同功能分组展示的能力，而对话框则当用户进行一些操作时，可以给出一些操作提示，这些都是非常实用的功能。本章将介绍用户界面中菜单和对话框的开发，同时还会对 Android 平台的 Toast 和 Notification 技术做简要介绍。

5.1　菜单 Menu

为了让 Android 应用程序有更加完美的用户体验，我们可以添加一些菜单来提示用户操作，让应用程序在功能上更完善。有时为了界面的美观，我们也可以把一些按钮用菜单的形式来表现。Android 平台下所提供的菜单分为 3 类，即选项菜单（Options Menu）、上下文菜单（Context Menu）和子菜单（Submenu）。

5.1.1　选项菜单 Options Menu 和子菜单 Submenu

不管在模拟器上，还是真机上，都有一个 Menu 键，单击该键就会弹出一个菜单，此菜单就是选项菜单。选项菜单的菜单项最多只能有 6 个，如果超过 6 个，系统会自动将最后一个菜单项显示为"更多"。如图 5.1 所示，单击 Menu 键，界面弹出菜单。

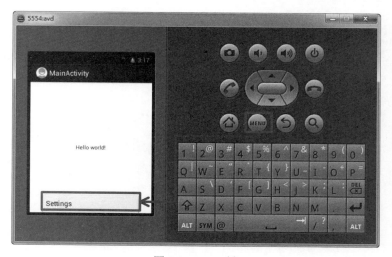

图 5.1　Menu 键

在 Android 中，通过回调方法来创建菜单并处理菜单按下的事件。这些回调方法及说明如表 5-1 所示。

表 5-1　选项菜单相关的回调方法及说明

方法名	描述
onCreateOptionsMenu(Menu menu)	初始化选项菜单，该方法只在第一次显示菜单时调用，如果需要每次显示菜单时更新菜单项，则需要重写 onPrepareOptionsMenu(Menu)方法
public boolean onOptionsItemSelected (MenuItem item)	当选项菜单中某个选项被选中时调用该方法，默认返回一个 false 的布尔值
public void onOptionsMenuClosed (Menu menu)	当选项菜单关闭时（或者由于用户按下了返回键，或者是选择了某个菜单选项）调用该方法
public boolean onPrepareOptionsMenu (Menu menu)	为程序准备选项菜单，每次选项菜单显示前会调用该方法。可以通过该方法设置某些菜单项可用或不可用，或者修改菜单项的内容。重写该方法时需要返回 true，否则选项菜单将不会显示

开发 Options Menu 主要涉及 Menu、MenuItem 和 Submenu，下面进行简单介绍。

1. Menu

一个 Menu 对象代表一个菜单，Menu 对象可以添加 MenuItem，也可以添加子菜单 Submenu。

2. MeniItem

MenuItem 对象代表一个菜单项，通常 MenuItem 实例通过 Menu.add()方法获得。

```
Menu.add (int groupId, int itemId, int order,CharSequence title);
```

其中，groupId 表示菜单项所在组 ID，itemId 表示菜单项 ID，order 表示菜单项顺序，title 表示菜单项显示的文本内容。

3. Submenu

每个 Submenu 实例代表一个子菜单。子菜单的添加是通过 Menu.addSubmenu()方法实现的。

【示例 5-1】 Options Menu 的使用。新建项目 OptionsMenu，在布局文件中添加一个 TextView 控件，用于显示被选择的菜单项文本内容，如图 5.2 所示。

控件	属性	值
TextView	id	@+id/textView1
	layout_width	wrap_content
	layout_heigth	wrap_content
	layout_centerHorizontal	ture
	layout_centerVertical	ture
	padding	@dimen/padding_medium
	text	@string/hello_world

图 5.2　OptionsMenu 界面图

逻辑代码如下：

```java
public class OptionsMenuActivity extends Activity {
    private static int FIRST = Menu.FIRST;
    private static int SECOND = Menu.FIRST+1;
    private TextView textView;
    @Override
    public void onCreate(Bundle savedInstanceState) {
        super.onCreate(savedInstanceState);
        setContentView(R.layout.activity_options_menu);
        textView = (TextView)findViewById(R.id.textView1);
    }
@Override
//重写onCreateOptionsMenu(Menu menu)方法,初始化选项菜单
    public boolean onCreateOptionsMenu(Menu menu) {
        menu.add(0,FIRST,1,"开始游戏");
        menu.add(0,SECOND,2,"暂停游戏");
        return super.onCreateOptionsMenu(menu);
    }
    @Override
//重写onOptionsItemSelected(MenuItem item)方法,通过判断itemId,修改TextView
//显示内容,为用户所选的对应菜单项内容
    public boolean onOptionsItemSelected(MenuItem item) {
     if(item.getItemId()==1){
        textView.setText("开始游戏");
        }
     if(item.getItemId()==2){
        textView.setText("暂停游戏");
        }
    return super.onOptionsItemSelected(item);
    }
}
```

运行程序，效果如图 5.3 所示。单击 Menu 键，弹出菜单项。选择不同的菜单项，界面显示对应的内容。

图 5.3 Options Menu

【示例 5-2】 在【示例 5-1】的基础上，调用 Menu.add(int groupId, int itemId, int order, CharSequence title)方法，添加菜单项"关于游戏"。然后为新添加的菜单项添加 OnMenuItemClickListener 监听器，处理菜单选中事件。

逻辑代码：

```java
public class OptionsMenuActivity extends Activity {
```

```
        private static int FIRST = Menu.FIRST;
        private static int SECOND = Menu.FIRST+1;
        //声明被添加菜单项的itemId
        private static int THREE= Menu.FIRST+2;
        private TextView textView;
    @Override
    public void onCreate(Bundle savedInstanceState) {
        super.onCreate(savedInstanceState);
        setContentView(R.layout.activity_options_menu);
        textView = (TextView)findViewById(R.id.textView1);
    }
    @Override
    public boolean onCreateOptionsMenu(Menu menu) {
        menu.add(0,FIRST,1,"开始游戏");
        menu.add(0,SECOND,2,"暂停游戏");
        //添加菜单项
        MenuItem item = menu.add(0, THREE, 3, "关于游戏");
        item.setOnMenuItemClickListener(new OnMenuItemClickListener() {
            //为菜单项添加OnMenuItemClickListener监听器,处理菜单选中事件,将TextView
文本显示为新添加的菜单项文本
            public boolean onMenuItemClick(MenuItem item) {
                // TODO Auto-generated method stub
                textView.setText("关于游戏");

                return false;
            }
        });

        return super.onCreateOptionsMenu(menu);
    }
}
```

运行程序，效果如图5.4所示。

图 5.4 添加菜单项

【示例 5-3】 在【示例 5-1】的基础上，添加子菜单"退出游戏"，并为其添加菜单项"确定"和"取消"。

逻辑代码如下：

```
@Override
    public boolean onCreateOptionsMenu(Menu menu) {
        menu.add(0,FIRST,1,"开始游戏");
        menu.add(0,SECOND,2,"暂停游戏");
```

```
    MenuItem item = menu.add(0, THREE, 3, "关于游戏");
   //添加子菜单
   final SubMenu subMenu=menu.addSubMenu(1, 100, 100, "退出游戏");
   //添加菜单项
   subMenu.add(2, 101, 101, "确定");
   subMenu.add(2, 102, 102, "取消");
    item.setOnMenuItemClickListener(new OnMenuItemClickListener() {
      public boolean onMenuItemClick(MenuItem item) {
         // TODO Auto-generated method stub
         textView.setText("关于游戏");

         return false;
      }
   });
```

运行程序，效果如图 5.5 所示。

图 5.5　添加子菜单

5.1.2　上下文菜单 Context Menu

在桌面平台中，上下文菜单即右键菜单，一般被绑定到指定的可视组件；在手机设备中，长按屏幕（触摸屏）或按压指定的功能按钮也会触发上下文菜单。使用上下文菜单时常用到 Activity 类的成员方法，如表 5-2 所示。

表 5-2　Activity 类中与 ContextMenu 相关的方法及说明

方 法 名 称	参 数 说 明	方 法 说 明
onCreateContextMenu (ContextMenu menu, View v, ContextMenu.ContextMenuInfo menuInfo)	menu：创建的上下文菜单； v：上下文菜单依附的View对象； menuInfo：上下文菜单需要额外显示的信息	每次为 View 对象呼出上下文菜单时都将调用该方法
onContextItemSelected(MenuItem item)	item：被选中的上下文菜单选项	当用户选择了上下文菜单选项后，调用该方法进行处理
onContextMenuClosed(Menu menu)	menu：被关闭的上下文菜单	当上下文菜单被关闭时调用该方法
registerForContextMenu (View view)	view：要显示上下文菜单的View对象	为指定的 View 对象注册一个上下文菜单

【示例 5-4】 Context Menu 的使用。新建一个项目 ContextMenu，在其布局文件中添加一个 Button 控件。然后在逻辑代码部分为 Button 注册一个 Context Menu，并修改菜单头内容，如图 5.6 所示。

控件	属性	值
Button	id	@+id/tbutton1
	layout_width	wrap_content
	layout_heigth	wrap_content
	text	请选择一种出行方式

图 5.6 Context Menu1

逻辑代码如下：

```java
public class ContextMenuActivity extends Activity {
    private Button button;
    @Override
    public void onCreate(Bundle savedInstanceState) {
        super.onCreate(savedInstanceState);
        setContentView(R.layout.activity_context_menu);
        button = (Button)findViewById(R.id.button1);
        //为 Button 注册 ContextMenu
        registerForContextMenu(button);
    }
    @Override
    //每次为 View 对象呼出上下文菜单时调用该方法
    public void onCreateContextMenu(ContextMenu menu, View v,
            ContextMenuInfo menuInfo) {
        // TODO Auto-generated method stub
        super.onCreateContextMenu(menu, v, menuInfo);
            //设置 ContextMenu 的菜单头及菜单项
            if (v==button) {
            menu.setHeaderTitle("请选择一种出行方式");
            menu.add(200, 200, 200, "火车");
            menu.add(201, 201, 201, "飞机");
            }
    }
    @Override
    public boolean onContextItemSelected(MenuItem item) {
        // TODO Auto-generated method stub
            //通过判断 itemId 处理菜单选中事件修改 Button 文本内容
            if (item.getItemId()==200) {
                button.setText("火车");
            }else if (item.getItemId()==201) {
                button.setText("飞机");
            }
        return super.onContextItemSelected(item);
    }
```

}

运行程序,效果如图 5.7 所示。长按"请选择一种出行方式"按钮,呼出 Context Menu。选择不同的菜单项,按钮显示对应的文本内容。

图 5.7 Context Menu2

5.2 对话框 Dialog

与菜单界面一样,对话框也是应用程序常用的一种界面方式。对话框就是程序在运行时弹出的一个提示界面。这个提示页面可以通过不同形式的对话框来显示信息。Android 平台下的对话框,主要包括普通对话框、提示对话框、单选和复选对话框、列表对话框、进度对话框、日期与时间对话框等。

5.2.1 普通对话框 Dialog

本节介绍普通对话框的开发。普通对话框中只显示提示信息和一个确定按钮,通过 Dialog 来实现。

【示例 5-5】 普通对话框的使用。新建一个项目 Dialog,在其布局文件中添加一个 TextView 文本显示提示信息,再添加一个"确定"按钮。然后在逻辑代码部分创建一个 Dialog。

逻辑代码如下:

```java
public class DialogActivity extends Activity {
    @Override
    public void onCreate(Bundle savedInstanceState) {
        super.onCreate(savedInstanceState);
        //创建 Dialog 对象
        Dialog dialog=new Dialog(this);
        //加载布局文件,显示提示信息和确定按钮
        dialog.setContentView(R.layout.activity_dialog);
        //设置对话框标题
        dialog.setTitle("普通对话框");
        //显示对话框
        dialog.show();

    }
}
```

运行程序，效果如图 5.8 所示。

5.2.2 提示对话框 AlertDialog

AlertDialog 是一个提示对话框，它可以显示不同的内容，如列表、单选按钮、复选按钮等。AlertDialog 的构造方法被声明为 protected，所以不能直接使用 new 关键字来创建 AlertDialog 类的对象实例。要想创建 AlertDialog 对话框，需要使用 Builder 类，该类是 AlertDialog 类中定义的一个内嵌类。

【示例 5-6】 AlertDialog 的使用。新建项目 AlertDialog，创建一个带"确定"和"取消"按钮的 AlertDialog。

逻辑代码如下：

图 5.8 普通对话框

```
public class AlertDialogActivity extends Activity {
    @Override
    public void onCreate(Bundle savedInstanceState) {
        super.onCreate(savedInstanceState);
        //创建 AlertDialog 对象
        AlertDialog.Builder builder=new AlertDialog.Builder(this);
        //设置 AlertDialog 图标
        builder.setIcon(android.R.drawable.ic_dialog_info);
        //设置 AlertDialog 标题
        builder.setTitle("AlertDialog");
        //设置 AlertDialog 内容
        builder .setMessage("你确定删除吗？");
        //添加确定按钮
        builder.setPositiveButton("确定", new DialogInterface.OnClickListener(){
            public void onClick(DialogInterface dialog, int which) {
                setTitle("确定");
            }
        });
        //添加取消按钮
        builder.setNegativeButton("取消", new DialogInterface.OnClickListener(){
            public void onClick(DialogInterface dialog, int which) {
                setTitle("取消");
                }
        });
        //显示 AlertDialog
        builder.show();
    }
}
```

运行程序，效果如图 5.9 所示。

【示例 5-7】 创建单选按钮对话框。新建一个项目 RadioButtonDialog，并创建一个单选按钮对话框。

逻辑代码如下：

```
public class RadioButtonDialogActivity extends Activity {
    @Override
    public void onCreate(Bundle savedInstanceState) {
        super.onCreate(savedInstanceState);
        AlertDialog.Builder builder=new AlertDialog.Builder(this);
        builder.setIcon(android.R.drawable.ic_dialog_info);
```

```
        builder.setTitle("单选按钮对话框");
        //调用setSingleChoiceItems()方法为对话框设置单选按钮
        builder.setSingleChoiceItems(new String[] { "火车","飞机","轮船"}, 0,
            new DialogInterface.OnClickListener() {
          public void onClick(DialogInterface dialog, int which) {
              switch (which) {
              case 0:
                  break;
              default:
                  break;
              }
          }
        });
            builder.setPositiveButton("确定", null).show();
    }
}
```

运行程序,效果如图 5.10 所示。

图 5.9 AlertDialog

图 5.10 单选按钮对话框

【示例 5-8】 创建多选按钮对话框。新建一个项目 CheckBoxDialog,并创建一个多选按钮对话框。

逻辑代码如下:

```
public class CheckBoxDialogActivity extends Activity {
    @Override
    public void onCreate(Bundle savedInstanceState) {
        super.onCreate(savedInstanceState);
        AlertDialog.Builder builder=new AlertDialog.Builder(this);
        builder.setIcon(android.R.drawable.ic_dialog_info);
        builder.setTitle("多选按钮对话框");
        //调用setMultiChoiceItems()方法为对话框设置多选按钮
        builder.setMultiChoiceItems( new String[] {"火车","飞机","轮船"},
        null, null);
        builder.setPositiveButton("确定", null);
        builder.setNegativeButton("取消", null).show();
    }
}
```

运行程序，效果如图5.11所示。

图5.11 多选按钮对话框

【示例5-9】 创建列表对话框。新建一个项目ListDialog，并创建一个列表对话框。逻辑代码如下：

```java
public class ListDialogActivity extends Activity {
    @Override
    public void onCreate(Bundle savedInstanceState) {
        super.onCreate(savedInstanceState);
        AlertDialog.Builder builder=new AlertDialog.Builder(this);
        builder.setIcon(android.R.drawable.ic_dialog_info);
        builder.setTitle("列表对话框");
        //调用setItems()方法，为对话框设置列表
        builder.setItems(new String[] {"火车","飞机","轮船"}, null);
        builder.setPositiveButton("确定", null);
        builder.show();
    }
}
```

运行程序，效果如图5.12所示。

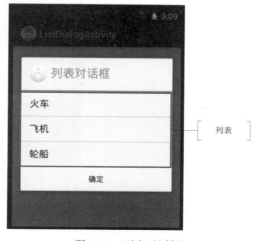

图5.12 列表对话框

5.2.3 进度对话框 ProgressDialog

进度对话框（ProgressDialog）能够给用户一个进度的提示，如在下载时，可以显示下载的进度；在加载时，也可以采用进度条对话框。ProgressDialog 通过调用 setProgressStyle() 方法，可以设置显示圆形进度样式，也可以显示水平进度样式。

❑ ProgressDialog.STYLE_HORIZONTAL：水平进度样式。
❑ ProgressDialog.STYLE_SPINNER：圆形进度样式。

【示例 5-10】 进度对话框。新建项目 ProgressDialog，设置不同的进度条样式，显示不同的进度对话框。

逻辑代码如下：

```java
public class ProgressDialogActivity extends Activity {
    @Override
    public void onCreate(Bundle savedInstanceState) {
        super.onCreate(savedInstanceState);
        //创建 ProgressDialog 对象
        ProgressDialog dialog = new ProgressDialog(this);
        //设置对话框标题
        dialog.setTitle("进度对话框");
        //设置对话框内容
        dialog.setMessage("请稍等...");
        //设置进度条样式
        dialog.setProgressStyle( ProgressDialog.STYLE_SPINNER );
        //显示对话框
        dialog.show();
    }
}
```

运行程序，效果如图 5.13 所示。

图 5.13 圆形进度对话框

图 5.14 水平进度对话框

修改进度条为水平样式。运行程序，效果如图 5.14 所示。

```java
dialog.setProgressStyle( ProgressDialog.STYLE_HORIZONTAL);
```

5.2.4 日期选择对话框 DatePickerDialog

日期选择对话框 DatePickerDialog，就是在对话框中显示日期，并且用户可以修改日期。

【示例 5-11】 DatePickerDialog 的使用。新建一个项目 DatePickerDialog，调用 Calendar 类的静态方法 getInstance()，初始化一个 Calendar 对象，然后使用 Calendar 对象获取系统日期。创建一个 DatePickerDialog 显示系统日期。

逻辑代码如下：

```java
public class DatePickerDialogActivity extends Activity {
    @Override
    public void onCreate(Bundle savedInstanceState) {
        super.onCreate(savedInstanceState);
        setContentView(R.layout.activity_date_picker_dialog);
        //初始化 Calendar 对象
        Calendar calendar = Calendar.getInstance();
        //创建 DatePickerDialog 对象
        DatePickerDialog dialog = new DatePickerDialog(this, null,
        //获取系统日期,传入 DatePickerDialog 对象
            calendar.get(Calendar.YEAR),
            calendar.get(Calendar.MONTH),
            calendar.get(Calendar.DAY_OF_MONTH));
        //显示 DatePickerDialog
        dialog.show();
    }
}
```

运行程序，效果如图 5.15 所示。用户可以通过滑动当前日期上下的浅色日期，进行选择修改。

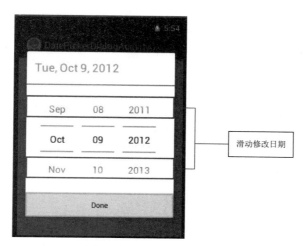

图 5.15　DatePickerDialog

5.2.5 时间选择对话框 TimePickerDialog

时间选择对话框 TimePickerDialog，就是在对话框中显示时间，用户可以修改时间。

【示例 5-12】 TimePickerDialog 的使用。新建一个项目 TimePickerDialog，调用 Calendar 类的静态方法 getInstance()，初始化一个日历对象。然后使用 Calendar 对象获取系统时间。

创建一个 TimePickerDialog 显示系统时间。

逻辑代码如下：

```
public class TimePickerDialogActivity extends Activity {
    @Override
    public void onCreate(Bundle savedInstanceState) {
        super.onCreate(savedInstanceState);
        setContentView(R.layout.activity_time_picker_dialog);
        //初始化 Calendar 对象
        Calendar calendar = Calendar .getInstance();
        //创建 TimePickerDialog 对象
        TimePickerDialog dialog = new TimePickerDialog(this, null,
            //获取系统时间,传入 TimePickerDialog 对象
            calendar.get(Calendar.HOUR_OF_DAY),
            calendar.get(Calendar.MINUTE),
            //设置显示格式为 24 小时制
            true);
        //显示 DatePickerDialog
        dialog.show();
    }
}
```

运行程序，效果如图 5.16 所示。用户可以通过滑动当前日期上下的浅色时间，进行选择修改。

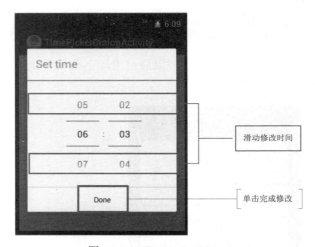

图 5.16　TimePickerDialog

5.3　Android 中的温馨提示

在 Android 中有两个提示信息的控件。一个是 Toast，它默认显示在界面的底部，做一些简单的提示；另一个是 Notification，用过 Android 手机的读者都知道，在有未接电话或有新短信未读取时，在标题栏中就会有相应的图标进行提示，这就是 Notification 控件。

5.3.1　消息提示条 Toast

Toast 是一种非常方便的消息提示框,它向用户提示比较快速的即时消息。相比对话框,

消息提示条没有焦点，且显示时长有限，显示后间隔一段时间会自动消失。Toast 的用法比较简单，步骤介绍如下：

（1）调用 Toast.makeText()方法；

（2）设置方法中的参数：上下文环境、Toast 显示的提示消息、Toast 的显示时长（时长的参数有两种，其中 Toast.LENGTH_LONG 表示长显示；Toast.LENGTH_SHORT 表示短显示）；

（3）调用 show()方法，显示 Toast。

【示例 5-13】Toast 的使用。新建一个项目 Toast，在布局文件中添加一个 Button 按钮。在逻辑代码部分为 Button 注册监听，然后使用 Toast 提示监听注册成功。

逻辑代码如下：

```
public class ToastActivity extends Activity {
    @Override
    public void onCreate(Bundle savedInstanceState) {
        super.onCreate(savedInstanceState);
        setContentView(R.layout.activity_toast);
        Button button = (Button)findViewById(R.id.button1);
        button.setOnClickListener(new OnClickListener() {
            public void onClick(View v) {
                // TODO Auto-generated method stub
                //调用 Toast.makeText()方法
                Toast.makeText(
                //当前上下文环境
                ToastActivity.this,
                // Toast 显示内容
                "注册成功",
                // Toast 显示时长
                Toast.LENGTH_LONG)
                //显示 Toast
                .show();
            }
        });
    }
}
```

运行程序，效果如图 5.17 所示。

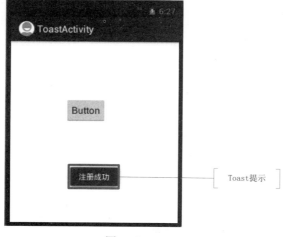

图 5.17　Toast

5.3.2 通知 Notification

Notification 可以在屏幕最顶部的状态栏中显示一个图标通知。用手指按下状态栏，并从手机上方向下滑动，就可以打开状态栏查看提示消息。通知的同时，可以播放声音及振动提示用户。单击通知还可以进入指定的 Activity，如图 5.18 所示。

图 5.18　Notification 示意图

开发 Notification，主要涉及以下三个类。

- Notification.Builder：一般用于动态地设置 Notification 的一些属性，即用 set 来设置；
- NotificationManager：主要负责将 Notification 在状态栏中显示和取消；
- Notification：主要是设置 Notification 的相关属性。

下面给出 Notification 类中的一些常量，如表 5-3 所示。

表 5-3　Notification 的常量

常　　量	说　　明
DEFAULT_ALL	使用所有默认值，比如声音、振动、闪屏等
DEFAULT_LIGHTS	使用默认闪光提示
DEFAULT_SOUNDS	使用默认提示声音
DEFAULT_VIBRATE	使用默认手机振动提示

注意：使用默认手机振动提示，需要在 manifest.xml 中加入权限：<uses-permission android:name="android.permission.VIBRATE" />。但是手机振动在模拟器上没有效果，最好在 Android 真机进行测试。

【示例 5-14】　Notification 的使用。新建一个项目 Notification，在布局文件 activity_notification.xml 中，添加一个按钮 click。单击按钮，显示通知。

逻辑代码如下：

```
public class NotificationActivity extends Activity {
    //声明 Notification
    private Notification notification;
    //声明 NotificationManager
```

```java
    private NotificationManager nManager;
    //声明 Notification.Builder
    private Notification.Builder nBuilder;
    //定义 Notification 的 id
    private int notification_id=11;
    //声明 Button
    private Button click;
    @Override
    public void onCreate(Bundle savedInstanceState) {
        super.onCreate(savedInstanceState);
        setContentView(R.layout.activity_notification);
        click = (Button)findViewById(R.id.button);
        click.setOnClickListener(new OnClickListener() {

            public void onClick(View v) {
                // TODO Auto-generated method stub
                //通过获取系统服务来获取 NotificationManager 对象
                nManager = (NotificationManager)getSystemService
                (NOTIFICATION_SERVICE);
                // new 一个 Notification.Builder 对象
                nBuilder = new Notification.Builder
                (NotificationActivity.this);
                //设置 Notification 的效果为默认闪光提示
                nBuilder.setDefaults(Notification.DEFAULT_LIGHTS);
                //设置 Notification 第一次出现在状态栏时的文本
                nBuilder.setTicker("A new notification");
                //设置 Notification 的大标题
                nBuilder.setContentTitle("Notification");
                //设置 Notification 的小标题
                nBuilder.setContentText("you hava a new message");
                //设置 Notification 的图标
                nBuilder.setSmallIcon(R.drawable.ic_launcher);
                //创建 Intent 对象，作为 PendingIntent 参数
                Intent intent = new Intent(NotificationActivity.this,
                NotificationActivity.class);
                //创建一个与 Activity 相关联的 PendingIntent 对象
                PendingIntent pi = PendingIntent.getActivity
                (NotificationActivity.this, 0, intent, 0);
                //在单击状态栏上的通知时就会打开所关联的 Activity,
                    并通过 intent 把参数带到新 Activity
                nBuilder.setContentIntent(pi);
                //创建通知
                notification = nBuilder.build();
                //发送通知
                nManager.notify(notification_id, notification);

            }
        });
    }
}
```

运行程序，效果如图 5.19 所示。单击 click 按钮，状态栏显示通知。

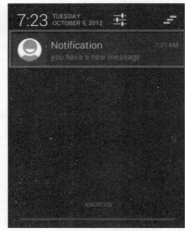

图 5.19 Notification

5.4 小　　结

本章主要介绍了 Android 中一些辅助功能的实现。主要包括菜单、对话框、Toast 和 Notification。其中，菜单、对话框和 Toast 都比较简单。Notification 在开发时，涉及的类比较多，是本章难点。需要读者认真学习，熟练掌握。

5.5 习　　题

1. 新建项目 OptionsMenu。添加两个 OptionsMenu："开始游戏"和"暂停游戏"，在单击"开始游戏"按钮时，使用 Toast 提示。然后添加一个菜单项"关于游戏"，和一个子菜单"退出游戏"，程序运行效果如图 5.20 所示。

图 5.20 OptionsMenu

【分析】本题目综合考查了读者对 OptionsMenu 的掌握，添加菜单项、添加子菜单，以及使用 Toast 显示提示信息。在开发过程中，覆盖 onCreateOptionsMenu(Menu menu)方法初始化菜单项，覆盖 onOptionsItemSelected(MenuItem item)方法处理选中事件。

【核心代码】本题的核心代码如下所示。
逻辑代码：

```java
@Override
public boolean onCreateOptionsMenu(Menu menu) {

    menu.add(0,FIRST,1,"开始游戏");
    menu.add(0,SECOND,2,"暂停游戏");
    MenuItem item = menu.add(0, FOUR, 4, "关于游戏");

    final SubMenu subMenu=menu.addSubMenu(1, 100, 100, "退出游戏");
    subMenu.add(2, 101, 101, "确定");
    subMenu.add(2, 102, 102, "取消");

    item.setOnMenuItemClickListener(new OnMenuItemClickListener() {

      public boolean onMenuItemClick(MenuItem item) {
          // TODO Auto-generated method stub
          textView.setText("关于游戏");

          return false;
      }
  });


    return super.onCreateOptionsMenu(menu);
}

 @Override
 public boolean onOptionsItemSelected(MenuItem item) {
   if(item.getItemId()==1){
      Toast.makeText(OptionsMenuActivity.this,"开始游戏",Toast.
      LENGTH_LONG).show();
      }
       if(item.getItemId()==2){
          textView.setText("暂停游戏");
          }
       if(item.getItemId()==3){
          textView.setText("退出游戏");
          }
  return super.onOptionsItemSelected(item);
}
```

2. 新建项目 ContextMenu。在界面添加一个按钮，为其注册 ContextMenu。选择喜欢的铃声，如图 5.21 所示。

图 5.21 ContextMenu

【分析】本题目考查了读者对 ContextMenu 的掌握。调用 registerForContextMenu(View view)方法，将 ContextMenu 注册到按钮上使用。然后重写 onCreateContextMenu(ContextMenu View,ContextMenu.ContextMenuInfo)方法，对按钮进行相应修改。

【核心代码】本题的核心代码如下所示。

```java
public void onCreate(Bundle savedInstanceState) {
    super.onCreate(savedInstanceState);
    setContentView(R.layout.activity_context_menu);
    button = (Button)findViewById(R.id.button1);
    registerForContextMenu(button);
}

@Override
public void onCreateContextMenu(ContextMenu menu, View v,
        ContextMenuInfo menuInfo) {
    // TODO Auto-generated method stub
    super.onCreateContextMenu(menu, v, menuInfo);
        if (v==button) {
        menu.setHeaderTitle("请选择一种响铃方式");
        menu.add(200, 200, 200, "铃声");
        menu.add(201, 201, 201, "振动");
        }
}

@Override
public boolean onContextItemSelected(MenuItem item) {
    // TODO Auto-generated method stub
        if (item.getItemId()==200) {
        button.setText("铃声");
        }else if (item.getItemId()==201) {
        button.setText("振动");
        }
    return super.onContextItemSelected(item);
}
```

3. 新建项目 AlertDialog，运行程序显示提示对话框，如图 5.22 所示。

【分析】本题目考查了读者对 AlertDialog 的掌握。创建 AlertDialog 对话框，需要使用 Builder 类，而不能直接使用 new 实例化一个对象出来。然后设置对话框的图标、标题和内容。最后调用 Builder 类的 setPositiveButton()方法和 setNegativeButton()方法添加"确定"和"取消"按钮。

【核心代码】本题的核心代码如下所示。

```java
AlertDialog.Builder builder=new AlertDialog.Builder(this);
builder.setIcon(android.R.drawable.ic_dialog_info);
builder.setTitle("AlertDialog");
builder.setMessage("确定更新？");
builder.setPositiveButton("确定", new DialogInterface.
OnClickListener(){
        public void onClick(DialogInterface dialog, int which) {
            setTitle("确定");
        }
});

builder.setNegativeButton("取消", new DialogInterface.
OnClickListener() {
```

```
        public void onClick(DialogInterface dialog, int which) {
            setTitle("取消");
        }
});

builder.show();
```

图 5.22　AlertDialog

图 5.23　RadioButtonDialog

4. 新建项目 RadioButtonDialog，运行程序显示"单选按钮对话框"，如图 5.23 所示。

【分析】本题目考查了读者对单选按钮对话框的掌握。使用 Builder 类，创建 AlertDialog 对话框。然后调用 setSingleChoiceItems()方法，为对话框设置单选按钮。调用 setPositiveButton()方法，为 AlertDialog 添加"确定"按钮。最后调用 show()方法来显示对话框。

【核心代码】本题的核心代码如下所示。

逻辑代码：

```
AlertDialog.Builder builder=new AlertDialog.Builder(this);
builder.setIcon(android.R.drawable.ic_dialog_info);
builder.setTitle("单选按钮对话框");
builder.setSingleChoiceItems(new String[] { "响铃","振动","静音"}, 0,
    new DialogInterface.OnClickListener() {
    public void onClick(DialogInterface dialog, int which) {
        switch (which) {
        case 0:
            break;
        default:
            break;
        }
    }
});

builder.setPositiveButton("确定", null).show();
```

5. 新建项目 CheckBoxDialog，运行程序显示"多选按钮对话框"，如图 5.24 所示。

【分析】本题目考查了读者对多选按钮对话框的掌握。使用 Builder 类，创建 AlertDialog 对话框。然后调用 setMultiChoiceItems()方法，为对话框设置多选按钮。调用 setPositiveButton()方法，为 AlertDialog 添加"确定"按钮。调用 setNegativeButton()方法，添加"取消"按钮。最后调用 show()方法来显示对话框。

【核心代码】本题的核心代码如下所示。
逻辑代码：

```
AlertDialog.Builder builder=new AlertDialog.Builder(this);
builder.setIcon(android.R.drawable.ic_dialog_info);
builder.setTitle("多选按钮对话框");
builder.setMultiChoiceItems( new String[] {"响铃","振动","静音"},
null, null);
builder.setPositiveButton("确定", null);
builder.setNegativeButton("取消", null).show();
```

图 5.24　CheckBoxDialog

图 5.25　ListDialog

6. 新建项目 ListDialog，运行程序显示"列表对话框"，如图 5.25 所示。

【分析】本题目考查了读者对列表对话框的掌握。使用 Builder 类，创建 AlertDialog 对话框。然后调用 setItems()方法，为对话框设置列表。调用 setPositiveButton()方法，为 AlertDialog 添加"确定"按钮。最后调用 show()方法来显示对话框。

【核心代码】本题的核心代码如下所示。
逻辑代码：

```
AlertDialog.Builder builder=new AlertDialog.Builder(this);
builder.setIcon(android.R.drawable.ic_dialog_info);
builder.setTitle("列表对话框");
builder.setItems(new String[] { "响铃","振动","静音"}, null);
builder.setPositiveButton("确定", null);
builder.show();
```

第 6 章　Activity 和 Intent

在第一章中我们了解到，Activity 和 Intent 是 Android 中的两大组件。Activity 是程序的界面，负责与用户交互。Intent 是一个消息框架，可以在不同组件之间传递数据信息。下面我们将深入学习 Activity 和 Intent。

6.1　Activity 生命周期

Activty 的生命周期，也就是它所在进程的生命周期。Android 系统将所有的进程，大致分为以下 5 类进行管理。

- 前台进程：即当前正在前台运行的进程，说明用户当前正通过该进程与系统进行交互，所以该进程为最重要的进程，除非系统的内存已经到不堪重负的情况，否则系统是不会将该进程中止的。
- 可见进程：一般显示在屏幕中，但是用户并没有直接与之进行交互。例如，某个应用程序运行时，根据用户的操作正在显示某个对话框，此时对话框后面的进程便为可见进程，该进程对用户来说同样是非常重要的进程，除非为了保证前台进程的正常运行，否则 Android 系统一般是不会将该进程中止的。
- 服务进程：是拥有 Service 的进程，该进程一般是在后台为用户服务的，例如音乐播放器的播放、后台的任务管理等。一般情况下，Android 系统是不会将其中断的，除非系统的内存已经达到崩溃的边缘，必须通过释放该进程才能保证前台进程的正常运行时，才可能将其中止。
- 后台进程：一般对用户的作用不大，缺少该进程并不会影响用户对系统的体验。所以如果系统需要中止某个进程才能保证系统正常运行时，那么会有非常大的几率将该进程中止。
- 空进程：对用户没有任何作用的进程。该进程一般是为缓存机制服务的，当系统需要中止某个进程保证系统的正常服务时，会首先将该进程中止。

Activty 的生命周期是由 Android 系统来控制的。一般情况下，Android 系统会根据应用程序对用户的重要性，以及当前系统的负载，来决定生命周期的长短。

如图 6.1 所示是 Activity 的生命周期。矩形中的方法表示回调方法，当 Activity 状态转换时执行这些方法。椭圆中所示为 Activity 的主要状态。

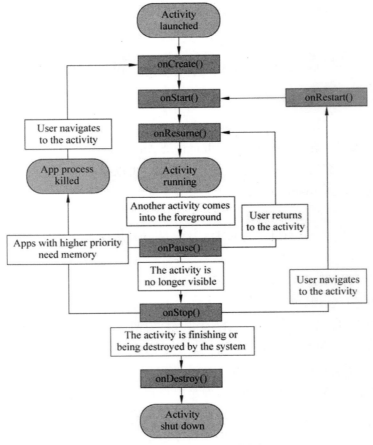

图 6.1 Activity 生命周期图

6.2 单界面程序

在 Android 应用程序中，一个 Activity 就是一个用户界面。用户与程序的交互，就是通过该类来实现的。

6.2.1 单界面程序启动

我们先从只有一个用户界面的 Android 程序，开始 Activity 的学习。

【示例 6-1】 创建一个简单的 Android 应用程序。该程序只包含一个 Activity 界面，界面中只有一个 OFF 按钮。运行程序，效果如图 6.2 所示。

位于前台的 Activity 总是处于活动状态。此时它是可视的、有焦点的。Android 系统会尽最大可能保持它的活动状态。如果系统资源不够该 Activity 运行，系统杀死其他 Activity，以确保它获得运行所需要的资源。

图 6.2 单界面程序

6.2.2 Activity 状态变化

每一个 Activity 都处于某一个状态。Activity 有 4 种基本状态：Active（活动）、Paused（暂停）、Stoped（停止）和 Killed（销毁）。对于开发者来说，是无法控制其应用程序处于某一个状态的，这些均由系统来完成。但是当一个活动的状态发生改变的时候，开发者可以通过调用 onXXX()方法，获取相关的通知信息。

【示例 6-2】 基于【示例 6-1】，通过覆盖 onStart()、onResume()方法，来了解 Activity 状态的具体变化。

逻辑代码如下：

```
@Override
//覆盖 onStart()方法
protected void onStart() {
// TODO Auto-generated method stub
 super.onStart();
 //执行 onStart()方法时,打印 onstart,以便查看 Activity 状态
  System.out.println("onStart");
}
@Override
//覆盖 onResume()方法
protected void onResume() {
// TODO Auto-generated method stub
 super.onResume();
 //执行 onResume ()方法时打印 onResume 以便查看 Activity 状态
  System.out.println("onResume");
}
```

运行程序，输出语句显示在 Eclipse 的 LogCat 面板中，如图 6.3 所示。Activity 状态变化如图 6.4 所示。

L...	Time	PID	TID	Application	Tag	Text
I	07-30 05:25:54.419	617	617	com.example.activity	System.out	onCreate
I	07-30 05:25:54.430	617	617	com.example.activity	System.out	onStart
I	07-30 05:25:54.430	617	617	com.example.activity	System.out	onResume

图 6.3 输出语句

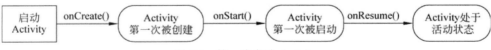

图 6.4 第一次启动 Activity

第一次启动 MainActivity，依次执行以下方法：onCreate()、MainActivity created、onStart()、MainActivity started、onResume()、MainActivity actived，进入活动状态。

注意：Logcat 是 Android SDK 中的一个通用日志工具。在程序的运行过程中可以通过 Logcat 打印状态信息和错误信息等。Logcat 另外一个重要的用途是在程序启动和初始化的过程中向开发者报告进展状况。Logcat 打开的方法：在 Eclipse 中依次选择 windows、show view、other、android、logcat 命令即可。

6.2.3 单界面程序退出

调用 finish()方法，可以结束当前正在运行的 Activity。MainActivity 从暂停到停止，再到销毁，彻底被杀死后退出程序。

【示例 6-3】 为 OFF 按钮绑定监听。在监听中调用 finish()方法结束 Activity。应用程序退出，返回到 HOME 界面。

逻辑代码如下：

```java
@Override
    public void onCreate(Bundle savedInstanceState) {
        super.onCreate(savedInstanceState);
        setContentView(R.layout.activity_main);
        System.out.println("onCreate");

        Button offButton = (Button)findViewById(R.id.button1);
        offButton.setOnClickListener(new OnClickListener() {

            public void onClick(View v) {
                // TODO Auto-generated method stub
                //调用 finish()方法结束 Activity
                finish();
            }
        });
    }
@Override
protected void onPause() {
// TODO Auto-generated method stub
 super.onPause();
  //执行 onPause()方法时打印 onPause,以便查看 Activity 状态
  System.out.println("onPause");
}

@Override
protected void onStop() {
// TODO Auto-generated method stub
 super.onStop();
 //执行 onStop()方法时打印 onStop,以便查看 Activity 状态
 System.out.println("onStop");
}
@Override
protected void onRestart() {
// TODO Auto-generated method stub
 super.onRestart();
  //执行 onRestart()方法时打印 onRestart,以便查看 Activity 状态
  System.out.println("onRestart");
}
@Override
protected void onDestroy() {
// TODO Auto-generated method stub
 super.onDestroy();
  //执行 onDestory()方法时打印 onDestory,以便查看 Activity 状态
  System.out.println("onDestory");
}



}
```

运行程序，单击按钮触发监听事件，查看 LogCat 面板中输出的信息，如图 6.5 所示。Activity 状态变化如图 6.6 所示。

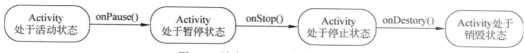

图 6.5　LogCat 面板中输出的信息

```
Activity          onPause()    Activity          onStop()    Activity          onDestory()    Activity处于
处于活动状态  ─────────→  处于暂停状态  ─────────→  处于停止状态  ──────────→   销毁状态
```

图 6.6　结束 Activity 状态变化图

结束 MainActivity，依次执行以下方法：onPause()、MainActivitypaused、onStop()、MainActivity stoped、onDestory ()、MainActivity killed。

首先，Activity 由活动状态转为暂停状态。此时它依然与窗口管理器保持连接，系统继续维护其内部状态，所以它仍然可见。但它已经失去了焦点，故不可与用户交互。在极特殊的情况下，Android 将会杀死一个暂停的 Activity，来为活动的 Activity 提供充足的资源。

接着该 Activity 被停止，变为完全隐藏、失去焦点，并且不可见。但是，系统将仍然在内存中保存它所有的状态和信息。

最后该 Activity 被杀死，转为销毁状态。Activity 结束，退出当前应用程序。

MainActivity 结束，返回 HOME 界面。运行效果如图 6.7 所示。

图 6.7　返回 HOME 界面

注：由于该项目运行，是模拟器首次启动，所以单击按钮，结束当前 Activity，返回到 HOME 界面。如果模拟器不是首次启动，单击 OFF 按钮，则会返回到上一次运行的应用程序界面。

6.3　多界面程序

一个应用程序通常有不同的界面，由此可知一个 Android 应用程序可以由多个 Activity 组成。接下来我们将以包含两个 Activity 的应用程序为例，学习两个 Activity 如何实现跳转。

6.3.1　启动第一个 Activity——主 Activity

主 Activity 是程序启动的入口。应用程序成功启动之后，呈献给用户的第一个界面，即为该程序的主 Activity。

【示例 6-4】　新建一个项目 ActivityLifeCylce，在其布局文件中添加一个 Button 按钮。在逻辑代码部分，覆盖生命周期方法。运行程序，启动 Activity，查看 Activity 的状态变化。

逻辑代码如下：

```java
@Override
protected void onStart() {                                  //覆盖onStart()方法
// TODO Auto-generated method stub
 super.onStart();
  System.out.println("onStart");
}
@Override
protected void onResume() {                                 //覆盖onResume()方法
// TODO Auto-generated method stub
 super.onResume();
  System.out.println("onResume");
}
@Override
protected void onPause() {                                  //覆盖onPause()方法
// TODO Auto-generated method stub
 super.onPause();
  System.out.println("onPause");
}
@Override
protected void onStop() {                                   //覆盖onStop()方法
// TODO Auto-generated method stub
 super.onStop();
  System.out.println("onStop");
}
@Override
protected void onRestart() {                                //覆盖onRestart()方法
// TODO Auto-generated method stub
 super.onRestart();
  System.out.println("onRestart");
}
@Override
protected void onDestroy() {                                //覆盖onDestroy()方法
// TODO Auto-generated method stub
 super.onDestroy();
  System.out.println("onDestory");
}
```

运行程序，查看 LogCat 面板中输出的信息，如图 6.8 所示。

L...	Time	PID	TID	Application	Tag	Text
I	07-30 08:23:57.645	679	679	com.example.activity1...	System.out	onCreate
I	07-30 08:23:57.645	679	679	com.example.activity1...	System.out	onStart
I	07-30 08:23:57.655	679	679	com.example.activity1...	System.out	onResume

图 6.8　LogCat 面板中输出的信息

第一次启动 ActivityLifeCylceActivity，依次执行以下方法：onCreate()、onStart()、onResume()，ActivityLifeCylceActivity 位于栈顶，处于活动状态。

启动 ActivityLifeCylceActivity，效果如图 6.9 所示。

6.3.2　新建第二个 Activity——Two

新建程序中的第二个 Activity 界面，以便实现界面的跳转。

【示例 6-5】 在【示例 6-4】的项目中新建一个 Activity——Two，与 ActivityLifeCylce-Activity 并列位于 ActivityLifeCylce 项目中。在其布局文件中添加一个按钮，在逻辑代码部

第 6 章 Activity 和 Intent

图 6.9　启动第一个 Activity

分覆盖生命周期方法。然后在 AndroidManifest.xml 文件中添加<activity>节点，声明 Two，与主 Activity 并列为<application>标签的直接子类（否则应用程序无法识别 Two）。

在 AndroidManifest.xml 文件中声明 Two：

```xml
<application
    android:icon="@drawable/ic_launcher"
    android:label="@string/app_name"
    android:theme="@style/AppTheme" >
    <activity
        android:name=".ActivityLifeCylceActivity"
        android:label="@string/title_activity_activity_life_cylce" >

        <intent-filter>
            <action android:name="android.intent.action.MAIN" />
            <category android:name="android.intent.category.LAUNCHER" />
        </intent-filter>
    </activity>
    <!--声明注册 Two -->
    <activity android:name="Two"></activity>

</application>
```

逻辑代码如下：

```java
@Override
protected void onStart() {                              //覆盖 onStart()方法
    // TODO Auto-generated method stub
    super.onStart();
    System.out.println("2onStart");
}

@Override
protected void onResume() {                             //覆盖 onResume()方法
    // TODO Auto-generated method stub
    super.onResume();
    System.out.println("2onResume");
}

@Override
protected void onPause() {                              //覆盖 onPause()方法
    // TODO Auto-generated method stub
```

```java
    super.onPause();
    System.out.println("2onPause");
}

@Override
protected void onStop() {                          //覆盖 onStop()方法
    // TODO Auto-generated method stub
    super.onStop();
    System.out.println("2onStop");
}

@Override
protected void onRestart() {                       //覆盖 onRestart()方法
    // TODO Auto-generated method stub
    super.onRestart();
    System.out.println("2onRestart");
}

@Override
protected void onDestroy() {                       //覆盖 onDestroy()方法
    // TODO Auto-generated method stub
    super.onDestroy();
    System.out.println("2onDestory");
}
```

6.3.3 启动 Two

为【示例 6-4】中主 Activity 的 Button 绑定监听。在监听中声明 Intent（详细介绍见 6.5 节）对象，调用 setClass()方法确定目标组件，然后调用 startActivity()方法启动 Two。触发主 Activity 的按钮单击监听事件，跳转到 Two 界面。

逻辑代码如下：

```java
Button button1 = (Button)findViewById(R.id.button1);
button1.setOnClickListener(new OnClickListener() {
    public void onClick(View v) {
        // TODO Auto-generated method stub
        //创建 Intent 对象
        Intent intent = new Intent();
        //确定目标组件
        intent.setClass(ActivityLifeCylceActivity.this, Two.class);
        //启动目标组件
        startActivity(intent);
    }
});
```

运行程序，查看 LogCat 面板中输出的信息，如图 6.10 所示。Activity 状态变化如图 6.11 所示。

I	07-30 08:30:21.926	679	679	com.example.activityl...	System.out	onPause
I	07-30 08:30:22.295	679	679	com.example.activityl...	System.out	2onCreate
I	07-30 08:30:22.295	679	679	com.example.activityl...	System.out	2onStart
I	07-30 08:30:22.295	679	679	com.example.activityl...	System.out	2onResume

图 6.10　LogCat 面板中输出的信息

图 6.11 启动第二个 Activity

启动第二个 Activity 时，第一个 Activity onPause()转为暂停状态，因为第二个 Activity 需要在前台运行。这时候需要将活动的状态持久化，比如正在编辑的数据库记录等。

第二个 Activity 第一次启动，依次执行 onCreate()、onStart()、onResume()方法。Two 位于栈顶，处于活动状态。

第一个 Activity 不再需要展示给用户，执行 onStop()方法，ActivityLifeCylceActivity 被压入栈底，转为停止状态。

注意：如果内存紧张，系统会直接结束这个活动，而不会触发 onStop() 方法。所以保存状态信息是应该在 onPause 时做，而不是在 onStop 时做。活动如果没有在前台运行，都将被停止，或者是 Linux 管理进程为了给新的活动预留足够的存储空间而随时结束这些活动。因此对于开发者来说，在设计应用程序的时候，必须时刻牢记这一原则。在一些情况下，onPause()方法或许是活动触发的最后的方法，因此开发者需要在这个时候保存需要保存的信息。

Two 被启动，效果如图 6.12 所示。

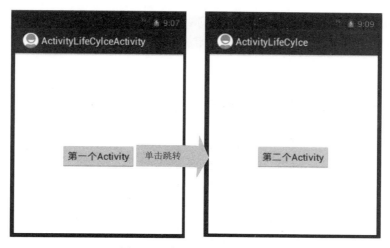

图 6.12 启动第二个 Activity

6.3.4 跳转回主 Activity

为 Two 界面中的按钮绑定监听，在监听中声明 Intent 对象。然后指定主 Activity 为目标组件，触发按钮单击监听事件，调用 startActivity()方法实现跳转回主 Activity。

逻辑代码如下：

```
        Button button2 = (Button)findViewById(R.id.button2);
        button2.setOnClickListener(new OnClickListener() {
            public void onClick(View v) {
                // TODO Auto-generated method stub
                Intent intent = new Intent();
```

```
                intent.setClass(Two.this, ActivityLifeCylceActivity.class);
                startActivity(intent);
            }
        });
    }
```

运行程序，查看 LogCat 面板中输出的信息，如图 6.13 所示。Activity 状态变化如图 6.14 所示。

图 6.13　LogCat 面板中输出的信息

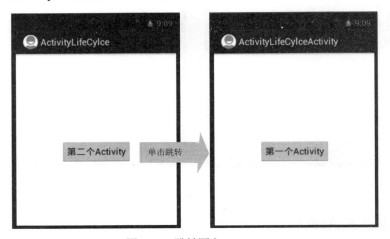

图 6.14　跳转回主 Activity

当返回到第一个 Activity 时，第二个 Activity onPause() 转为暂停状态。

第一个 Activity onRestart()被重新启动，再次展现给用户，然后执行 onStart()、onResume()方法。ActivityLifeCylceActivity 再次位于栈顶，由停止状态转为活动状态。第二个 Activity 执行 onStop()方法，被压入栈底，转为停止状态。

跳转回主 Activity，效果如图 6.15 所示。

图 6.15　跳转回主 Activity

6.3.5　BACK 到第二个 Activity

当用户单击手机上的 BACK 键时，结果和调用 Activity.finish()方法一样：告诉 Activity Manager，该 Activity 实例完成了相应的工作，可以被"回收"。

跳转回主 Activity 之后，单击 BACK 键，查看 LogCat 面板中输出的信息如图 6.16 所示。Activity 状态变化如图 6.17 所示。

图 6.16　LogCat 面板中输出的信息

点击 BACK 键 —onPause()→ 第一个 Activity 暂停 —2onRestart() 2onStart() 2onResume()→ 第二个 Activity 处于活动状态 —onStop()→ 第一个 Activity 停止 —onDestory()→ 第一个 Activity 销毁

图 6.17　BACK 到第二个 Activity

单击 BACK 键后，第一个 Activity onPause()转为停止状态；第二个 Activity onRestart()，被重新启动，再次展现给用户，然后执行 onStart()、onResume()方法；第一个 Activity 依次执行 onStop()、onDestory()方法，转为销毁状态。

单击 BACK 键之后，又跳转到第二个 Activity 界面，效果如图 6.18 所示。

图 6.18　BACK 到第二个 Activity

6.4　两个 Activity 之间传递数据

从上一节内容中，我们学习了两个 Activity 之间通过 Intent 类实现互相跳转。那么，在实现跳转的同时，我们能不能在 Activity 之间传递数据呢？答案是肯定的。下面我们将介绍 Activity 之间数据的传递。

6.4.1　传递数据到目标 Activity

其实，实现数据的传递很简单。只要在创建 Intent 对象后，为 Intent 对象绑定数据。当调用 startActivity(intent)方法启动目标组件时，Intent 就可以将数据从当前 Activity 传递到目标组件。语法格式如下：

```
Intent intent=new Intent();              // 声明 Intent 对象
intent.setClass(……,                      // 当前 Activity
```

```
                ......)                      // 目标 Activity
Bundle bundle = new Bundle();                // 声明 Bundle 对象
bundle.putString("","");                     // 设置 Bundle 内容
intent.putExtras(bundle);                    // 绑定 Bundle 到 Intent
startActivity(intent);                       // 启动目标 Activity
```

【示例 6-6】 新建一个项目 TransferData，在布局文件中添加一个 Tranfer 按钮，再新建一个 GetDataActivity 作为目标组件。单击 Tranfer 按钮，传递数据到 GetDataActivity。在 GetDataActivity 中，声明一个 Intent 对象，然后通过 getExtras()方法来获取数据。

TransferDataActivity 逻辑代码如下：

```
public class TransferDataActivity extends Activity {
    @Override
    public void onCreate(Bundle savedInstanceState) {
        super.onCreate(savedInstanceState);
        setContentView(R.layout.activity_transfer_data);
        Button button1 = (Button)findViewById(R.id.button1);
        button1.setOnClickListener(new OnClickListener() {
            public void onClick(View v) {
                // TODO Auto-generated method stub
                Intent intent=new Intent();
                intent.setClass(TransferDataActivity.this,
                GetDataActivity.class);
                //声明 Bundle 对像
                Bundle bundle = new Bundle();
                //设置 Bundle 内容
                bundle.putString("data", "100");
                //绑定 Bundle 到 Intent
                intent.putExtras(bundle);
                startActivity(intent);
            }
        });
    }
}
```

GetDataActivity 逻辑代码如下;

```
public class GetDataActivity extends Activity {
    @Override
    public void onCreate(Bundle savedInstanceState) {
        super.onCreate(savedInstanceState);
        setContentView(R.layout.activity_get_data);
        //声明 Intent 对像
        Intent intent=new Intent();
        //获取数据
        String result = intent.getStringExtra("data");
        // Toast 显示获取的数据
        Toast.makeText(GetDataActivity.this, result, Toast.LENGTH_SHORT).
        show();
    }
}
```

运行程序，效果如图 6.19 所示。

第 6 章 Activity 和 Intent

图 6.19 主 Activity 传递数据到目标组件

6.4.2 返回数据到主 Activity

既然我们可以将数据从主 Activity 传递到另一个目标 Activity。那么，当我们期望在结束目标 Activity 时，获得它的返回结果，我们怎样将该数据再返回到主 Activity？下面通过案例具体演示。

【示例 6-7】 目标 Activity 把接收到的数据返回到主 Activity。新建一个项目 Extra，在布局文件中添加按钮，单击按钮，传递数据到目标组件 ResultActivity。在目标组件 ResultActivity 中也添加按钮，单击该按钮，将主 Activity 传递的数据返回。

ExtraActivity 逻辑代码如下：

```java
public class ExtraActivity extends Activity {
    @Override
    public void onCreate(Bundle savedInstanceState) {
        super.onCreate(savedInstanceState);
        setContentView(R.layout.activity_extra);
        Button button1 = (Button)findViewById(R.id.button1);
        button1.setOnClickListener(new OnClickListener() {
                public void onClick(View v) {
                    // TODO Auto-generated method stub
                    Intent intent=new Intent();
                    intent.setClass(ExtraActivity.this, ResultActivity.class);
                    Bundle bundle = new Bundle();
                    bundle.putString("data", "1");
                    intent.putExtras(bundle);
                    //调用 startActivityForResult(Intent intent, int requestCode)
                      方法启动目标组件
                    startActivityForResult(intent,0);
                }
        });
    }
    //调用 onActivityResult(int requestCode, int resultCode, intent data)
      方法通过判断结果码获得返回值
    @Override
    protected void onActivityResult(int requestCode, int resultCode, Intent data){
        // TODO Auto-generated method stub
```

```
    super.onActivityResult(requestCode, resultCode, data);
    //结果码匹配成功 Toast 显示返回值
    switch (resultCode) {
        case RESULT_OK:
            String str = data.getExtras().getString("data");
            Toast.makeText(this, str, Toast.LENGTH_SHORT).show();
            break;
        default:
            break;
    }
}
```

ResultActivity 逻辑代码如下：

```
public class ResultActivity extends Activity {
    @Override
    public void onCreate(Bundle savedInstanceState) {
        super.onCreate(savedInstanceState);
        setContentView(R.layout.activity_result);
        final Intent intent = new Intent();
        String result = intent.getStringExtra("data");
        Toast.makeText(ResultActivity.this, result, Toast.LENGTH_SHORT).show();
        Button button2 = (Button)findViewById(R.id.button2);
        button2.setOnClickListener(new OnClickListener() {
            public void onClick(View v) {
                // TODO Auto-generated method stub
                //返回数据到主 Activity
                setResult(RESULT_OK, intent);
                //结束当前 Activity 并返回主 Activity
                finish();
            }
        });
    }
}
```

在 AndroidManifest.xml 文件中声明 ResultActivity：

```xml
<application
    android:icon="@drawable/ic_launcher"
    android:label="@string/app_name"
    android:theme="@style/AppTheme" >
    <activity
        android:name=".ExtraActivity"
        android:label="@string/title_activity_extra" >
        <intent-filter>
            <action android:name="android.intent.action.MAIN" />

            <category android:name="android.intent.category.LAUNCHER" />
        </intent-filter>
    </activity>
    <!--声明注册 ResultActivity -->
    <activity android:name="ResultActivity"></activity>

</application>
```

运行程序，效果如图 6.20 所示。

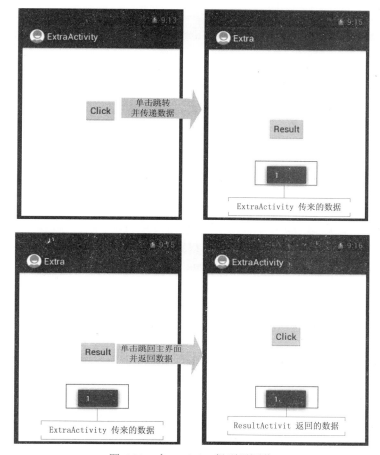

图 6.20　主 Activity 得到返回值

6.5　Intent 和 IntentFilter

在以上示例中，我们学习到 Activity 的跳转是通过 Intent 类实现的。Intent 代表了 Android 应用的启动"意图"，Android 会根据 Intent 的 setClass()方法的第二个参数来指定即将启动的目标组件。除此之外，我们还可以通过设置 Intent 的各个属性来启动目标组件。

6.5.1　意图 Intent

Intent 由 ComponentName、Action、Data、Category、Extra 及 Flag6 部分组成，Intent 通过调用 setXXX()方法来设置对应属性。下面将分别对其进行详细介绍。

1. ComponentName

ComponentName 对象，用于标识唯一的应用程序组件，即指明了期望的 Intent 组件。这种对象的名称是由目标组件的包名与目标组件的类名组合而成。在 Intent 传递过程中，组件名称是一个可选项。当指定它时，便是显式的 Intent 消息；而当不指定它时，Android 系统则会根据其他信息，以及 IntentFilter 的过滤条件选择相应的组件。

ComponentName 使用语法如下：

```
ComponentName cName = new ComponentName(    // 声明一个 ComponentName 对象
        "……",                                // 目标 Activity 所在包名
        "……");                               // 目标 Activity 所在包名+类名
intent.setComponent(cName);                  // 设置 Intent 的 ComponentName
```

2. Action

Action 实际上就是一个描述了 Intent 所触发动作名称的字符串。在 Intent 类中，已经定义好很多字符串常量来表示不同的 Action。当然，开发人员也可以自定义 Action，其定义规则同样非常简单。

系统定义的 Action 常量有很多，下面只列出其中一些较常见的。

- ACTION_CALL：拨出 Data 里封装的电话号码。
- ACTION_EDIT：打开 Data 里指定数据所对应的编辑程序。
- ACTION_VIEW：打开能够显示 Data 之中封装的数据的应用程序。
- ACTION_MAIN：声明程序的入口，该 Action 并不会接收任何数据，同时结束后也不会返回任何数据。
- ACTION_BOOT_COMPLETED：BroadcastReceiver Action 的常量，表明系统启动完毕。
- ACTION_TIME_CHANGED：BroadcastReceiver Action 的常量，表示系统时间通过设置而改变。

3. Data

Data 主要是对 Intent 消息中数据的封装，主要描述 Intent 的动作所操作到的数据的 URI 及类型。不同类型的 Action 会有不同的 Data 封装，例如，打电话的 Intent 会封装 tel://格式的电话 URI，而 ACTION_VIEW 的 Intent 中 Data 则会封装 http://格式的 URI。正确的 Data 封装对 Intent 匹配请求同样非常重要。

Data 使用语法如下：

```
Uri uri = Uri.parse(" ");       // 解析给定的 URI 字符编码为 Uri 对象
intent.setData(uri);            // 设置 Intent 的 Data 属性
```

4. Category

Category 是对目标组件类别信息的描述，为一个字符串对象。一个 Intent 中可以包含多个 Category，与 Category 相关的方法有 3 个，addCategory 是添加一个 Category、removeCategory 是删除一个 Category、getCategories 是得到一个 Category。Android 系统同样定义了一组静态字符常量来表示 Intent 的不同类别，下面列出一些常见的 Category 常量。

- CATEGORY_GADGET：表示目标 Activity 是可以嵌入到其他 Activity 中的。
- CATEGORY_HOME：表明目标 Activity 为 HOME Activity。
- CATEGORY_TAB：表明目标 Activity 是 TabActivity 的一个标签下的 Activity。
- CATEGORY_LAUNCHER：表明目标 Activity 是应用程序中最先被执行的 Activity。
- CATEGORY_PREFERNCE：表明目标 Activity 是一个首选的 Activity。

5. Extra

Extra 中封装了一些额外的附加信息，这些信息是认键值对的形式存在的。Intent 可以通过 putExtras()与 getExtras()方法来存储和获取 Extra。Android 系统的 Intent 类中，同样对一些常用的 Extra 键值进行了定义，下面列出一些常用的定义。

- EXTRA_CC：邮件抄送人邮箱地址。
- EXTRA_EMAIL：装有邮件发送地址的字符串数组。
- EXTRA_SUBJECT：当使用 ACTION_SEND 动作时，描述要发送邮件的主题。
- EXTRA_TEXT：当使用 ACTION_SEND 动作时，描述要发送文本的信息。

6. Flag

一些有关系统如何启动组件的标识位，Android 同样对其进行了封装。在开发程序中，一般不会用到。

【示例 6-8】 显式 Intent 的使用。新建一个项目 Intent，设置 Intent 的 ComponentName 属性，明确目标组件，实现界面跳转。

逻辑代码如下：

```java
Button button1 = (Button)findViewById(R.id.button1);
button1.setOnClickListener(new OnClickListener() {
    public void onClick(View v) {
        // TODO Auto-generated method stub
        //声明 Intent 对像
        Intent intent = new Intent();
        //声明 ComponentName 对像
        ComponentName cName = new ComponentName(
                // ComponentName 所在包名
                "com.example.intent",
                // ComponentName 所在包名及类名
                "com.example.intent.OneActivity");
        //显式的设置目标组件
        intent.setComponent(cName);
        //启动目标组件
        startActivity(intent);
    }
});
```

运行程序，触发 button1 的单击事件，跳转到 OneActivity 界面。运行效果如图 6.21 所示。

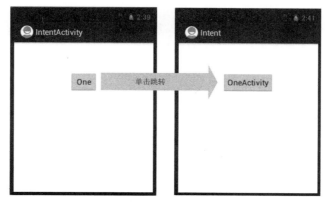

图 6.21 显式 Intent

【示例 6-9】 在【示例 6-8】主界面中，添加一个按钮 button2，并绑定监听。在监听中设置 Intent 的 Data 属性，打开 Data 内封装的数据所指代的程序。

逻辑代码如下：

```java
Button button2 = (Button)findViewById(R.id.button2);
button2.setOnClickListener(new OnClickListener() {
    public void onClick(View v) {
        // TODO Auto-generated method stub
        //声明 Intent 对像
        Intent intent = new Intent();
        //设置 Intent 执行动作打开 Data 中数据指代的程序
        intent.setAction(Intent.ACTION_VIEW);
        //解析该网址为 Uri 数据
        Uri uri = Uri.parse("http://www.google.com");
        //设置 Intent 操作的数据
        intent.setData(uri);
        //启动目标组件
        startActivity(intent);
    }
});
```

运行程序，触发 button2 的单击事件，运行效果如图 6.22 所示。

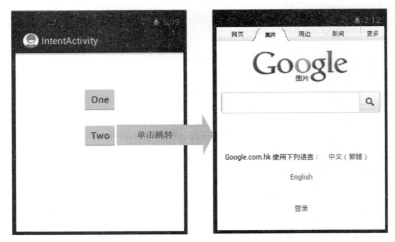

图 6.22　Data 属性使用

【示例 6-10】 在【示例 6-9】主界面中，添加 button3，并绑定监听。在监听中设置 Intent 的 Category 属性，明确目标组件，进入对应界面。

逻辑代码如下：

```java
Button button3 = (Button)findViewById(R.id.button3);
button3.setOnClickListener(new OnClickListener() {
    public void onClick(View v) {
        // TODO Auto-generated method stub
        //声明 Intent 对像
        Intent intent = new Intent();
        //声明程序入口
        intent.setAction(Intent.ACTION_MAIN);
```

```
            //设置目标组件为 HOME 界面
            intent.addCategory(Intent.CATEGORY_HOME);
            //启动目标组件
            startActivity(intent);
        }
    });
```

运行程序,触发 button3 的单击事件,运行效果如图 6.23 所示。

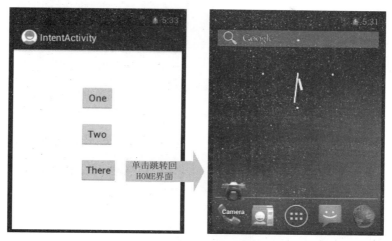

图 6.23　Category 属性使用

【示例 6-11】　在【示例 6-10】主界面中,添加 button4,并绑定监听。在监听中设置 Intent 的 Extra 属性,打开对应程序。

逻辑代码如下:

```
Button button4 = (Button)findViewById(R.id.button4);
button4.setOnClickListener(new OnClickListener() {
    public void onClick(View v) {
        // TODO Auto-generated method stub
        //接收人邮箱地址
        String recevier = "1065194059@qq.com";
        //抄送人邮箱地址
        String ccStrings = "113654201@163.com";
        //声明 Intent 对像
        Intent intent = new Intent();
        //添加接收人邮箱
        intent.putExtra(Intent.EXTRA_EMAIL, recevier);
        //添加抄送人邮箱
        intent.putExtra(Intent.EXTRA_CC, ccStrings);
        //添加邮件主题
        intent.putExtra(Intent.EXTRA_SUBJECT, "Theme");
        //添加邮件内容
        intent.putExtra(Intent.EXTRA_TEXT, "This is a test email");
        //启动目标组件
        startActivity(intent);

    }
});
```

运行程序,触发 button4 的单击事件,运行效果如图 6.24 所示。

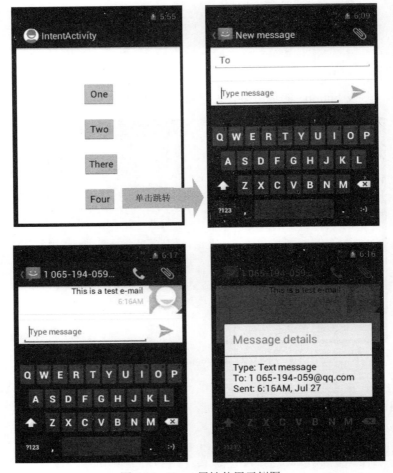

图 6.24　Extra 属性使用示例图

6.5.2　意图过滤器 IntentFilter

　　IntentFilter 描述了一个组件愿意接收什么样的 Intent 对象，Android 将其抽象为 android.content.IntentFilter 类。在 Android 的 AndroidManifest.xml 配置文件中可以通过 <intent-filter>节点，为一个 Activity 指定其 IntentFilter，以便告诉系统该 Activity 可以响应什么类型的 Intent。这样的 Intent 称为隐式的 Intent。

　　当程序员使用 startActivity(intent)来启动另外一个 Activity 时，如果直接指定了 intent 对象的 Component 属性，那么 Activity Manager 将试图启动其 Component 属性指定的 Activity。否则 Android 将通过 Intent 的其他属性，从安装在系统中的所有 Activity 中查找与之最匹配的一个启动。如果没有找到合适的 Activity，应用程序会得到一个系统抛出的异常。这个匹配的过程如图 6.25 所示。

1. 检查 Action

　　一个 Intent 只能设置一种 Action，但是一个 IntentFilter 却可以设置多个 Action 过滤。当 IntentFilter 设置了多个 Action 时,只需其中一个满足即可完成 Action 验证。当 IntentFilter

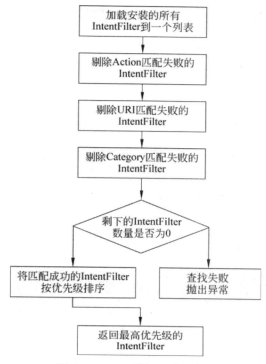

图 6.25　IntentFilter 匹配过程

中没有说明任何一个 Action 时，那么任何的 Action 都不会与之匹配，而如果 Intent 中没有包含任何 Action，那么只要 IntentFilter 中含有 Action 时，便会匹配成功。

2. 检查 Data

数据的监测主要包含两部分，即数据的 URI 和数据类型。而数据 URI 又被分成三部分进行匹配（scheme、authority、path），只有这些全部匹配时，Data 的验证才会成功。

3. 检查 Category

IntentFilter 同样可以设置多个 Category。当 Intent 中的 Category 与 IntentFilter 中的其中一个 Category 完全匹配时，便会通过 Category 的检查，而其他的 Category 并不受影响。但是当 IntentFilter 没有设置 Category 时，只能与没有设置 Category 的 Intent 相匹配。

【示例 6-12】　隐式 Intent 的使用。在【示例 6-11】中，添加一个按钮 button5，并绑定监听。在监听中添加代码，自定义 Action。在 AndroidManifest.xml 文件中，声明 TwoActivity，并添加<intent-filter>节点，设置其 action 和 category 属性。通过匹配 Action 和 Category 跳转到目标组件 TwoActivity。

逻辑代码如下：

```
Button button5 = (Button)findViewById(R.id.button5);
button5.setOnClickListener(new OnClickListener() {
    public void onClick(View v) {
        // TODO Auto-generated method stub
        //声明 Intent 对象
        Intent intent = new Intent();
```

```
            //设置自定义Action
            intent.setAction("com.example.intent.TwoActivity");
            //启动目标组件
            startActivity(intent);
        }
    });
```

AndroidManifest.xml 文件：

```xml
<activity android:name="TwoActivity">
    <intent-filter >
        <!--与自定义Action 一致匹配成功-->
        <action android:name="com.example.intent.TwoActivity"/>
        <!-- Category 为默认类型自动匹配-->
        <category android:name="android.intent.category.DEFAULT" />
    </intent-filter>
</activity>
```

运行程序，触发 button5 的单击事件，运行效果如图 6.26 所示。

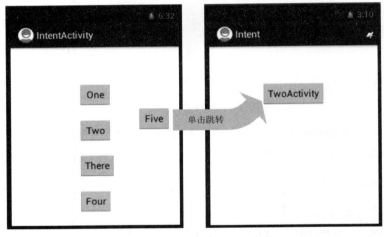

图 6.26　隐式 Intent

6.6　小　　结

本章详细介绍了 Android 的组件之一——Activity，从单界面到多界面的跳转，以及数据传递。本章重点就是学习如何开发 Activity、如何在 AndroidManifest.xml 文件中配置 Activity、掌握 Activity 生命周期等。本章难点在于实现多个 Activity 之间的数据共享，以及 Intent 和 IntentFilter 类。希望读者勤于练习，熟练掌握。

6.7　习　　题

新建项目 Extra，再新建一个 ResultActivity，实现两个 Activity 之间的切换的同时，传递数据到 ResultActivity，然后再由 ResultActivity 将数据返回到主 Activity，程序运行效果如图 6.27 所示。

图 6.27　返回数据到主 Activity

【分析】本题目考查了读者对 Activity 的掌握。在主 Activity 中调用 startActivityForResult(Intent intent, int requestCode)方法启动目标组件。在 ResultActivity 中，调用 setResult(int resultCode, intent data)方法，将传入的数据返回主 Activity，然后在主 Activity 中，调用 onActivityResult(int requestCode, int resultCode, intent data)方法获得返回值。

【核心代码】本题的核心代码如下所示。

ExtraActivity：

```
Button button1 = (Button)findViewById(R.id.button1);
button1.setOnClickListener(new OnClickListener() {
        public void onClick(View v) {
            // TODO Auto-generated method stub
            Intent intent=new Intent();
            intent.setClass(ExtraActivity.this, ResultActivity.class);
            Bundle bundle = new Bundle();
            bundle.putString("data", "9");
            intent.putExtras(bundle);
            startActivityForResult(intent,0);
        }
    });
@Override
protected void onActivityResult(int requestCode, int resultCode, Intent data) {
    // TODO Auto-generated method stub
    super.onActivityResult(requestCode, resultCode, data);
    switch (resultCode) {
       case RESULT_OK:

           String str = data.getExtras().getString("data");
           Toast.makeText(this, str, Toast.LENGTH_SHORT).show();
           break;
       default:
           break;
    }
}
```

ResultActivity：

```
        final Intent intent = getIntent();
        String result = intent.getStringExtra("data");
        Toast.makeText(ResultActivity.this, result, Toast.LENGTH_SHORT).
        show();
```

```java
Button button2 = (Button)findViewById(R.id.button2);
button2.setOnClickListener(new OnClickListener() {

    public void onClick(View v) {
        // TODO Auto-generated method stub
        setResult(RESULT_OK, intent);
        finish();
    }
});
```

第 7 章　Service 与 BroadcastReceiver

本章主要介绍 Android 平台中的另外两个组件，服务 Service 和消息广播接收者 BroadcastReceiver。Service 是一种后台运行的 Android 组件，BroadcastReceiver 是接受并响应广播通知的一类组件。本章将详细讲解 Service 的应用，并对 BroadcastReceiver 进行相应介绍。

7.1　Service 简介

Service 是 Android 系统中的重要组件，它在后台运行，能在后台加载数据、运行程序等。下面主要介绍 Service 的开发过程及它的生命周期。

7.1.1　Service 的特点和创建

Service 即"服务"，Service 在 Android 系统中占有很重要的位置。它具有以下几个特点：无法与用户直接进行交互；必须由用户或其他程序启动；优先级介于前台应用和后台应用之间。那么我们什么时候需要使用 Service 呢？例如，打开音乐播放之后又想打开电子书，并且不希望音乐停止播放，此时就可以使用 Service。

Service 就是在开启一个程序之后，又想打开另一个程序，同时不使前一个程序停止，仅仅将第一个程序转为后台运行时使用。Service 开发共分为两步：定义 Service 和配置 Service。下面依次讲解这两个步骤。

1．定义 Service

Android 提供了一个 Service 类。定义的时候，只要继承该类就可以了。定义的语法如下：

```
public class Service1 extends Service {       // 自定义 Service 子类继承于
                                              //   Service
    @Override
    public IBinder onBind(Intent intent) {    // 新建 Service 时系统自动覆盖
                                              //   onBind()方法,用于通信
        // TODO Auto-generated method stub
        return null;
    }
}
```

2．配置 Service

在 AndroidManifest.xml 文件中，配置该 Service。有以下两种配置方法。

第一种是显式配置，只需使用<Service…/>标签声明 Service 的名称。<Service…/>与其他组件标签（例如<activity></activity>）并列位于<application></application>标签内，为同

一个应用程序所用。

```
<service
android:name="Service1">         <!--Service 名称-->
</service>
```

第二种是隐式配置，除了声明 Service 名称之外，还需要为 Service 配置<intent-filter.../>子标签。通过匹配 Action 属性，说明该 Service 可以被哪些 Intent 启动。

```
<service android:name="Service1">
      <intent-filter>
         <!--设置 Action 属性-->
         <action android:name="android.service"/>
         <!--默认 Intent 类型-->
         <category android:name="android.intent.category.LAUNCHER" />
      </intent-filter>
</service>
```

7.1.2 Service 生命周期

Service 和 Activity 一样都有自己的生命周期，都是一个从创建到销毁的过程。Service 的启动方式有两种：context.startService()和 context.bindService()。使用 context.startService()启动 Service，访问者与 Service 没有关联，即使访问者退出了，Service 依然运行；使用 context.bindService()启动 Service，访问者与 Service 就绑定在一起，访问者一旦退出，Service 就终止运行，如图 7.1 所示。

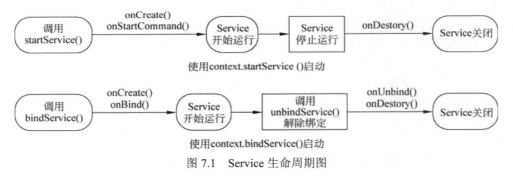

图 7.1 Service 生命周期图

下面给出 Service 生命周期中相关方法的说明，如表 7-1 所示。

表 7-1 Service 生命周期相关方法

方 法 名 称	方 法 说 明
startService(Intent service)	启动一个指定的应用程序服务
stopService(Intent service)	停止一个指定的应用程序服务
bindService(Intent, ServiceConnection, int)	连接到一个应用程序服务
unbindService(ServiceConnection conn)	从应用程序断开连接服务
onCreate()	第一次创建 Service 时执行该方法
onStartCommand()	每一次客户端通过调用 startService(Intent service)显式地启动服务时执行该方法

续表

方法名称	方法说明
onBind()	每一次客户端通过调用 bindService(Intent, ServiceConnection, int)隐式地启动服务时执行该方法绑定
onUnbind()	每个客户端断开与服务的绑定时执行该方法
onDestory()	当 Service 不再使用，并已经被删除时执行该方法

7.2 Service 操作

本节我们使用两种不同的方法启动 Service，观察各种情况下，Service 的生命周期以及状态变化。

7.2.1 使用 context.startService()启动 Service

Service 一般由 Activity 启动。当 Activity 调用 startService()方法启动 Service 时，如果 Service 还没有运行，则 Android 先调用 onCreate()，然后调用 onStartCommand()启动 Service，Service 进入运行状态。需要停止 Service 的时候直接使用 onDestroy()结束 Service。如果是调用者自己直接退出，而没有调用 stopService 的话，Service 会一直在后台运行。如图 7.2 所示。

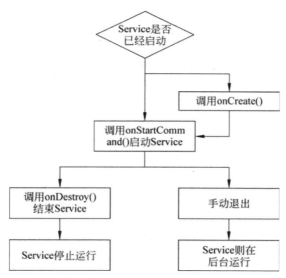

图 7.2 调用 startService()方法启动 Service 流程

【示例 7-1】 使用 context.startService()方法启动 Service。新建一个项目 Service1，在其布局文件中添加两个 Button。在 Service1Activity 中为两个按钮绑定监听，一个用于启动 Service，另一个用于停止 Service。创建 Service1 继承于 Service，在 AndroidManifest.xml 中显式配置 Service1。覆盖 Service 生命周期的相关方法，了解 Service 运行状态变化，输出提示信息在 LogCat 面板上。程序界面如图 7.3 所示。

控件	属性	值
Button	id	@+id/button1
	text	启动Service
Button	id	@+id/button2
	text	停止Service

图 7.3　Service1 界面

Service1：

```java
public class Service1 extends Service {
    @Override
    public void onCreate(){
        // TODO Auto-generated method stub
        //覆盖 onCreate()方法,执行该方法时打印 onCreate,以便查看 Service 状态
        super.onCreate();
        System.out.println("onCreate()");
    }
    @Override
    public int onStartCommand(Intent intent,int flags,int startId){
        // TODO Auto-generated method stub
        //覆盖 onStartCommand()方法,执行该方法时打印 onStartCommand,以便查看
          Service 状态
        System.out.println("onStartCommand()");
        return super.onStartCommand(intent,flags,startId);
    }

    @Override
    public void onDestroy(){
        // TODO Auto-generated method stub
        //覆盖 onDestory()方法,执行该方法时打印 onDestory,以便查看 Service 状态
        System.out.println("onDestroy()");
        super.onDestroy();
    }
    @Override
     public IBinder onBind(Intent intent) {
        // TODO Auto-generated method stub
        return null;
    }
}
```

Service1Activity：

```java
public class Service1Activity extends Activity {
    @Override
    public void onCreate(Bundle savedInstanceState) {
        super.onCreate(savedInstanceState);
        setContentView(R.layout.activity_service1);
        Button button1 = (Button)findViewById(R.id.button1);
        button1.setOnClickListener(new OnClickListener() {
```

```java
            public void onClick(View v) {
                // TODO Auto-generated method stub
                //调用startService()方法,启动Service
                startService();
            }
        });
        Button button2 = (Button)findViewById(R.id.button2);
        button2.setOnClickListener(new OnClickListener() {
            public void onClick(View v) {
                // TODO Auto-generated method stub
                //调用stopService()方法,停止Service
                stopService();
            }
        });
    }
    //创建stopService()方法停止Service
    protected void stopService() {
        // TODO Auto-generated method stub
        Intent intent =new Intent(this,Service1.class);
        stopService(intent);
    }
    //创建startService()方法启动Service
    protected void startService() {
        // TODO Auto-generated method stub
        Intent intent =new Intent(this,Service1.class);
        startService(intent);
    }
}
```

AndroidManifest.xml：

```xml
<application
    android:icon="@drawable/ic_launcher"
    android:label="@string/app_name"
    android:theme="@style/AppTheme" >
    <activity
        android:name=".Service1Activity"
        android:label="@string/title_activity_service1" >
        <intent-filter>
            <action android:name="android.intent.action.MAIN" />
            <category android:name="android.intent.category.LAUNCHER" />
        </intent-filter>
    </activity>

    <!--显式配置Service-- >
    <service android:name="Service1"></service>

</application>
```

运行程序，查看LogCat面板上输出的提示信息。

单击"启动Service"按钮,触发监听事件,Service依次执行onCreate()、onStartCommand()方法,如图7.4所示。

| I | 08-01 02:20:40.683 | 602 | 602 | com.example.service1 | System.out | onCreate() |
| I | 08-01 02:20:40.683 | 602 | 602 | com.example.service1 | System.out | onStartCommand() |

图7.4 单击"启动Service"按钮

单击"停止 Service"按钮，触发监听事件，Service 执行 onDestroy()方法，如图 7.5 所示。

```
I  08-01 02:22:27.663    602      602      com.example.service1    System.out    onDestroy()
```

图 7.5 单击"停止 Service"按钮

由【示例 7-1】可知，调用 context.startService()启动 Service 的过程为 onCreate()、onStartCommand()、onDestroy()。

7.2.2 使用 context.bindService()启动 Service

调用 context.bindService()方法，绑定 Service 到 Activity，依次执行 onCreate()和 onBind()方法，Service 被启动。调用 context.unBindService()解除绑定，Srevice 依次调用 onUnbind()和 onDestroy()方法退出服务。当结束与 Service 绑定的 Activity 时，Service 也会被终止。如图 7.6 所示。

图 7.6 调用 context.bindService()方法启动 Service 流程

【示例 7-2】使用 context.bindSerivce()方法启动 Service。新建一个项目 Service2，在其布局文件中添加两个 Button。在 Service2Activity 中为两个按钮绑定监听，一个用于启动 Service，另一个用于停止 Service。创建 Service2 继承于 Service，在 AndroidManifest.xml 中隐式配置 Service2。覆盖 Service 生命周期的相关方法，了解 Service 运行状态变化，输出提示信息在 LogCat 面板上。程序界面如图 7.7 所示。

图 7.7 Service2 界面

控件	属性	值
Button	id	@+id/button1
	text	启动Service
Button	id	@+id/button2
	text	停止Service

Service2

```
public class Service2 extends Service {
    @Override
    //覆盖 onCreate()方法,执行该方法时打印 onCreate,以便查看 Service 状态
    public void onCreate() {
        System.out.println("---onCreate---");
        super.onCreate();
    }
    @Override
```

```java
    //覆盖 onBind()方法,执行该方法时打印 onBind,以便查看 Service 状态
    public IBinder onBind(Intent intent) {
        // TODO Auto-generated method stub
        System.out.println("---onBind---");
        return null;
    }
    @Override
    //覆盖 onUnbind()方法,执行该方法时打印 onUnbind,以便查看 Service 状态
    public boolean onUnbind(Intent intent) {
        System.out.println("---onUnbind---");
        return super.onUnbind(intent);
    }
    @Override
    //覆盖 onDestory ()方法,执行该方法时打印 onDestory,以便查看 Service 状态
    public void onDestroy() {
        System.out.println("---onDestroy---");
        super.onDestroy();
    }
}
```

Service2Activity:

```java
public class Service2Activity extends Activity {
    //声明 Intent 对象,用于启动目标 Service
    private Intent intent = new Intent();
    //声明 ServiceConnection 对象,监听访问者与 Service 之间的连接情况
    private ServiceConnection sConnection = new ServiceConnection() {
        public void onServiceDisconnected(ComponentName name) {
            // TODO Auto-generated method stub
        }
        public void onServiceConnected(ComponentName name, IBinder service) {
            // TODO Auto-generated method stub
        }
    };
    @Override
    public void onCreate(Bundle savedInstanceState) {
        super.onCreate(savedInstanceState);
        setContentView(R.layout.activity_service2);
        //设置 Intent 所触发的 Action 字符串
        intent.setAction("android.intent.action.start");
        Button button1 = (Button)findViewById(R.id.button1);
        button1.setOnClickListener(new OnClickListener() {
            public void onClick(View v) {
                // TODO Auto-generated method stub
                //绑定 Service
                bindService(intent, sConnection, BIND_AUTO_CREATE);
            }
        });
        Button button2 = (Button)findViewById(R.id.button2);
        button2.setOnClickListener(new OnClickListener() {
            public void onClick(View v) {
                // TODO Auto-generated method stub
                //解除与 Service 的绑定
                unbindService(sConnection);
            }
        });
    }
}
```

AndroidManifest.xml：

```xml
<application
    android:icon="@drawable/ic_launcher"
    android:label="@string/app_name"
    android:theme="@style/AppTheme" >
    <activity
        android:name=".Service2Activity"
        android:label="@string/title_activity_service2" >
        <intent-filter>
            <action android:name="android.intent.action.MAIN" />

            <category android:name="android.intent.category.LAUNCHER" />
        </intent-filter>
    </activity>

    <service android:name="Service2">
        <intent-filter>
<!--隐式配置Service-->
            <action android:name="android.intent.action.start"/>
            <category android:name="android.intent.category.LAUNCHER" />
        </intent-filter>
    </service>
</application>
```

运行程序，查看 LogCat 面板输出的提示信息。

单击"启动 Service"按钮，触发监听事件，Service 依次执行 onCreate()、onBind()方法，如图 7.8 所示。

```
I  08-01 03:16:21.123   804   804   com.example.service2   System.out   ---onCreate---
I  08-01 03:16:21.133   804   804   com.example.service2   System.out   ---onBind---
```

图 7.8　单击"启动 Service"按钮

单击"停止 Service"按钮，触发监听事件，Service 执行 onUnbind()、onDestroy()方法，如图 7.9 所示。

```
I  08-01 03:17:06.543   804   804   com.example.service2   System.out   ---onUnbind---
I  08-01 03:17:06.543   804   804   com.example.service2   System.out   ---onDestroy---
```

图 7.9　单击"停止 Service"按钮

由示例 7-2 可知，调用 context.bindService()启动 Service 的过程为 onCreate()、onBind()、onUnbind()、onDestory()。其中，OnBind()只能被执行一次，不能多次执行。

7.3　Service 通信

根据通信方式，Service 可以分为两种类型：本地服务（Local Service）和远程服务（Remote Service）本地服务用于应用程序内部，远程服务用于 Android 系统内的应用程序之间。

7.3.1　本地服务通信

当程序通过 startService()和 stopService()方法启动和关闭 Service 时，Service 与访问者

之间没有太多关系。因此无法进行通信和数据交换。如果 Service 和访问者之间需要进行通信，应该调用 bindService()绑定 Service 与访问者，通信结束之后，再调用 unBindSevice()解除绑定，退出 Service。

绑定 Service 之后，Service 类中的 IBinder onbind(Intent intent) 方法的返回值，将传递给在访问者类中声明的 ServiceConnection 的 onServiceConnected(ComponentName name, IBinder service)方法中做参数。这样，访问者就可以通过 Inbind 对象，实现与 Service 之间的通信。交互关系如图 7.10 所示。

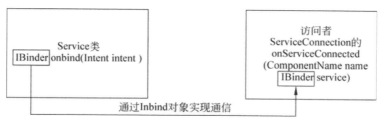

图 7.10 本地服务关系交互

context.bindSerivce(Intent service,ServiceConnection conn,int flags)方法语法如下：

```
context.bindSerivce(              // 绑定 Service
    Intent service,               // Intent 可以启动的目标 Service
    ServiceConnection conn,       /* ServiceConnection 对象,用于监听访问者与
                                     Service 之间的连接情况。
                                     当访问者与 Service 连接成功时,将回调
                                     ServiceConnection 对象的
                                     onServiceConnected()方法;如果断开将回调
                                     onServiceDisConnected()方法。*/
    int flags)
```

【示例 7-3】 本地服务与 Activity 通信。新建一个项目 Service3，在其布局文件中添加三个 Button。第一个用于启动 Service，第二个用于停止 Service，第三个用于获取数据。
Service3：

```java
public class Service3 extends Service {
    private int counter = 0;
    private boolean bRunning = true;
    //定义 onBinder 返回的对像
    private mBinder binder = new mBinder();
    //通过继承实现 Binder 类
    public class mBinder extends Binder{
        public int getCounter(){
            return counter;
        }
    }
    @Override
    public IBinder onBind(Intent arg0) {
        // TODO Auto-generated method stub
        //返回 IBinder 对像
        return binder;
    }

    @Override
    public void onCreate() {
```

```java
        // TODO Auto-generated method stub
        super.onCreate();
        //启动线程修改 counter 值
        new Thread(new Runnable() {
            public void run() {
                // TODO Auto-generated method stub
                while (bRunning = true) {
                    try {
                        Thread.sleep(1000);
                    } catch (InterruptedException e) {
                        // TODO Auto-generated catch block
                        e.printStackTrace();
                    }
                    counter ++;
                }
            }
        }).start();
    }
    @Override
    public boolean onUnbind(Intent intent) {
        // TODO Auto-generated method stub
        System.out.println("onUnbind");
        return true;
    }
    @Override
    public void onDestroy() {
        // TODO Auto-generated method stub
        super.onDestroy();
        bRunning = false;
        System.out.println("onDestroy");
    }
}
```

Service3Activity：

```java
public class Service3Activity extends Activity {
    private Intent intent = new Intent();
    //声明一个 Service3.mIbind 对象,用于获取数据
    private Service3.mBinder binder;
    private ServiceConnection sConnection = new ServiceConnection() {
        //解除绑定时输出 ServiceDisconnectd 提示
        public void onServiceDisconnected(ComponentName name) {
            // TODO Auto-generated method stub
            System.out.println("--ServiceDisconnected--");
            binder = null;
        }
        public void onServiceConnected(ComponentName name, IBinder service)
{
            // TODO Auto-generated method stub
            //绑定 Service 时输出 ServiceConnectd 提示
            System.out.println("--ServiceConnected--");
            //获取数据
            binder = (Service3.mBinder)service;
        }
    };
    @Override
    public void onCreate(Bundle savedInstanceState) {
        super.onCreate(savedInstanceState);
        setContentView(R.layout.activity_service3);
        intent.setAction("android.service");
```

```java
        Button button1 = (Button)findViewById(R.id.button1);
        button1.setOnClickListener(new OnClickListener() {
            public void onClick(View v) {
                // TODO Auto-generated method stub
                //绑定 Service
                bindService(intent, sConnection, BIND_AUTO_CREATE);
            }
        });
        Button button2 = (Button)findViewById(R.id.button2);
        button2.setOnClickListener(new OnClickListener() {
            public void onClick(View v) {
                // TODO Auto-generated method stub
                //解除绑定
                unbindService(sConnection);
            }
        });
        Button button3 = (Button)findViewById(R.id.button3);
        button3.setOnClickListener(new OnClickListener() {
            public void onClick(View v) {
                // TODO Auto-generated method stub
                // Toast 显示 counter 值
                Toast.makeText(Service3Activity.this,
                    "Service的counter值为" + binder.getCounter(),
                    Toast.LENGTH_LONG).show();
            }
        });
    }
}
```

运行程序，单击"启动 Service"按钮，然后单击"获取数据"按钮，出现 Toast 提示框，显示 counter 值。counter 由线程动态修改，每隔 1 秒钟，counter 的值加 1，效果如图 7.11 所示。

图 7.11 Service 通信

7.3.2 远程服务通信

在 Android 系统中，各应用程序都运行在自己的进程中。进程之间一般无法直接进行通信或者数据交换。Android 提供了 AIDL 工具来实现跨进程的通信。

AIDL（Android Interface Definition Language）是一种 IDL 接口定义语言((Interface Definition Language)，用于生成可以在 Android 设备上两个进程之间进行进程间通信（Internet Process Connection IPC）的代码。如果在一个进程中（例如 Activity）要调用另一个进程中（例如 Service）的对象，就可以使用 AIDL 来实现。

【示例 7-4】 使用 AIDL 实现跨进程通信。

1. 创建.aidl 文件

新建一个项目 MyAIDL，右击 src 项目包名，选择 new|File 命令，弹出如下对话框。新建文件 MyAIDL.aidl，单击"确定"按钮，如图 7.12 所示。

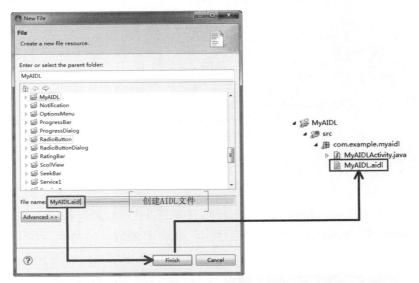

图 7.12 创建 AIDL 文件

2. 定义接口

生成的接口包含一个名为 Stub 的抽象的内部类，该类声明了所有.aidl 中描述的方法。Stub 还定义了少量的辅助方法，尤其是 asInterface()，通过它可以获得 IBinder（当 applicationContext.bindService()成功调用时传递到客户端的 onServiceConnected()），并且返回用于调用 IPC 方法的接口实例对象。

MyAIDL.aidl:

```
//项目包名
package com.example.myaidl;
    //定义接口
    interface MyAIDL{
        String getValue();
    }
```

保存 MyAIDL.aidl 文件， gen 目录下自动生成以 MyAIDL 命名的.java 接口文件——MyAIDL.java。如图 7.13 所示。

注意：接口的定义必须正确，否则无法自动生成 MyAIDL.java 接口文件。

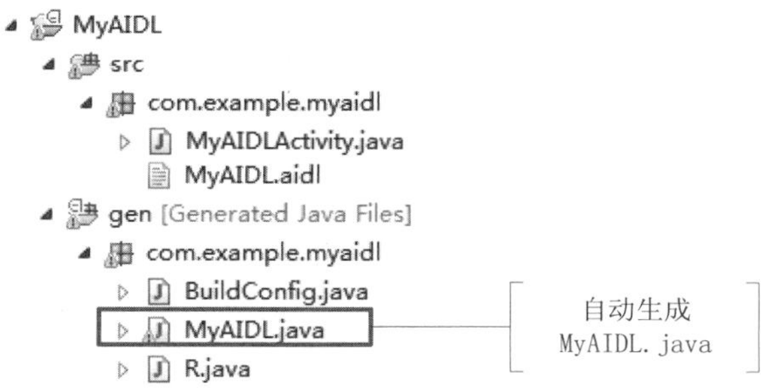

第 7 章 Service 与 BroadcastReceiver

图 7.13 自动生成 MyAIDL.java

MyAIDLService：

```java
public class MyAIDLService extends Service {
    //声明字符串数组
    private String[] values = {
            "java",
            "c++",
            "Android"
    };
    //数组下标
    private int index = 0;
    private boolean bRunning = true;
    //实现 MyAIDL 接口
    class mBinder extends MyAIDL.Stub{
        public String getValue() throws RemoteException {
            // TODO Auto-generated method stub
            return values[index];
        }
    }
    @Override
    public IBinder onBind(Intent intent) {
        // TODO Auto-generated method stub
        //返回 IBinder 实例对象
        return new mBinder();
    }
    @Override
    public void onCreate() {
        // TODO Auto-generated method stub
        super.onCreate();
        new Thread(new Runnable() {
            //启动线程随机产生 index 值
            public void run() {
                // TODO Auto-generated method stub
                while(bRunning = true){
                    index = (int)(Math.random()*2);
                    try {
                        Thread.sleep(1000);
                    } catch (InterruptedException e) {
                        // TODO Auto-generated catch block
                        e.printStackTrace();
                    }
                }
            }
        }).start();
    }
    @Override
    public boolean onUnbind(Intent intent) {
```

```java
        // TODO Auto-generated method stub
        return super.onUnbind(intent);
    }
    @Override
    public void onDestroy() {
        // TODO Auto-generated method stub
        super.onDestroy();
    }
}
```

MyAIDLActivity：

```java
public class MyAIDLActivity extends Activity {
//声明.aidl 文件中定义的接口类型的变量
    private MyAIDL myAIDL;
    private Intent intent;
    private ServiceConnection sConnection = new ServiceConnection() {
        public void onServiceDisconnected(ComponentName name) {
            // TODO Auto-generated method stub
        }
        public void onServiceConnected(ComponentName name, IBinder service) {
            // TODO Auto-generated method stub
            //将返回值转换为 MyAIDL 类型
            myAIDL = MyAIDL.Stub.asInterface(service);
        }
    };
    @Override
    public void onCreate(Bundle savedInstanceState) {
        super.onCreate(savedInstanceState);
        setContentView(R.layout.activity_my_aidl);
        intent = new Intent(getApplicationContext(),MyAIDLService.class);
    }
    @Override
    //响应 onKeyDown()处理事件
    public boolean onKeyDown(int keyCode, KeyEvent event) {
        //按向下方向键，绑定 Service
        if (keyCode == KeyEvent.KEYCODE_DPAD_DOWN) {
            bindService(intent, sConnection, BIND_AUTO_CREATE);
        //按向上方向键解除绑定
        }else if (keyCode == KeyEvent.KEYCODE_DPAD_UP) {
            unbindService(sConnection);
        //按中间键调用接口方法获取 value 值
        }else if (keyCode == KeyEvent.KEYCODE_DPAD_CENTER) {
            try {
                System.out.println("您选择了：" + myAIDL.getValue());
            } catch (RemoteException e) {
                // TODO Auto-generated catch block
                e.printStackTrace();
            }
        }
        return true;
    }
}
```

运行程序，在 LogCat 面板中查看效果，如图 7.14 所示。

```
I   08-02 08:03:07.352   1568   1568   com.example.myaidl   System.out   您选择了：java
I   08-02 08:03:09.402   1568   1568   com.example.myaidl   System.out   您选择了：c++
```

图 7.14　LogCat 面板中输出的信息

7.4 系统 Service

不管是智能手机还是非智能手机，Android 作为手机操作系统中的一员，在其系统 Service 中提供了最基本的打电话、发短信的服务供用户使用。这些服务主要通过 getSystemService(String serviceName)方法获得。本节将介绍一些常用的系统 Service。

7.4.1 电话管理器 TelephonyManager

TelephonyManager 是 Android 提供的系统服务之一。通过它可以获取手机的各种相关信息，例如设备 ID 号、SIM 卡号等。该 Service 使用流程介绍如下。

1. 获取 TELEPHONY_SERVICE 系统服务

使用该 Service 之前，首先需要获取系统服务。语法如下：

```
TelephonyManager telephonyManager;          // 声明 telephonyManager 对象
telephonyManager = (TelephonyManager)getSystemService(TELEPHONY_SERVICE);
                                            // 获取 TELEPHONY_SERVICE 系统服务
```

因为 getSystemService(String name)方法返回值类型是 Object，所以需要强制转换类型为 TelephonyManager 类。

2. 获取 TelephonyManager 相关属性

通过 TelephonyManager 对象，用户可以获取 Android 关于电话的多项属性信息，如电话类型、电话号码等。其语法形式如下：

```
String phoneType = "phoneType" + telephonyManager.getPhoneType();
                                            // 调用对应方法，获得相关信息
```

其中，TelephonyManager 常见的属性如表 7-2 所示。

表 7-2 TelephonyManager 常见属性表

方法名称	返回值类型	返回值说明
getPhonyType()	int	返回电话类型
getLineNumber()	String	返回电话号码
getDeviceId()	String	返回设备 ID
getSimNumber()	String	返回 sim 卡号
getNetworkOperatorName()	String	返回网络注册运营商名称

3. 调用 TelephonyManager 相关方法

TelephonyManager 类中，提供了与电话相关的方法，如表 7-3 所示。

表 7-3 TelephonyManager 相关方法

方法名称	方法说明
public boolean isNetworkRoaming()	如果设备漫游在目前网络，返回 true
public void listen(PhoneStateListener listener, int events)	注册监听对象，响应电话状态改变事件

【示例 7-5】 使用 TELEPHONY_SERVICE 系统服务。新建项目 TelephonyManager，TELEPHONY_SERVICE 系统服务。通过调用相关方法，获取电话相关信息。然后添加"获取信息"按钮，触发按钮单击监听事件，输出信息在 Logcat 面板中。

逻辑代码如下：

```java
public class TelephonyManagerActivity extends Activity {
//声明 TelephonyManager 对象
    private TelephonyManager telephonyManager;
    @Override
    public void onCreate(Bundle savedInstanceState) {
        super.onCreate(savedInstanceState);
        setContentView(R.layout.activity_telephony_manager);
        //获取 TELEPHONY_SERVICE 服务
        telephonyManager = (TelephonyManager)getSystemService
        (TELEPHONY_SERVICE);
        //获取电话类型
        final String phoneType = "phoneType" + telephonyManager.
getPhoneType();
        //获取电话号码
        final String phoneNumber = "phoneNumber" + telephonyManager.
getLine1Number();
        //获取设备 id
        final String deviceId = "deviceId" + telephonyManager.getDeviceId();
        //获取 sim 卡号
        final String simNumber = "simNumber" + telephonyManager.
getSimSerialNumber();
        //获取网络注册运营商名称
        final String netWorkCountry = "netWorkCountry" + telephonyManager.
getNetworkOperatorName();
        Button button = (Button)findViewById(R.id.button1);
        button.setOnClickListener(new OnClickListener() {
            public void onClick(View v) {
                // TODO Auto-generated method stub
                System.out.println(phoneType + "\n" + phoneNumber + "\n"
                    + deviceId + "\n" + simNumber + "\n" + netWorkCountry);
            }
        });
    }
}
```

使用 TelephonyManager 系统服务，需要在 AndroidManifest.xml 中添加用户权限"android.permission.READ_PHONE_STATE"。打开 AndroidManifest.xml 的 permission 面板开始添加，步骤如图 7.15 所示。

切换至 AndroidManifest.xml 面板查看，权限添加成功。

```xml
<manifest xmlns:android="http://schemas.android.com/apk/res/android"
    package="com.example.telephonymanager"
    android:versionCode="1"
    android:versionName="1.0" >
    <uses-sdk
        android:minSdkVersion="8"
        android:targetSdkVersion="15" />

    <!--添加用户权限，允许读取电话状态-->
    <uses-permission android:name="android.permission.READ_PHONE_STATE"/>
```

第 7 章 Service 与 BroadcastReceiver

图 7.15 添加用户权限

```
<application
    android:icon="@drawable/ic_launcher"
    android:label="@string/app_name"
    android:theme="@style/AppTheme" >
    <activity
        android:name=".TelephonyManagerActivity"
        android:label="@string/title_activity_telephony_manager" >
        <intent-filter>
            <action android:name="android.intent.action.MAIN" />
            <category android:name="android.intent.category.LAUNCHER" />
        </intent-filter>
    </activity>
</application>
</manifest>
```

提示：开发者也可查阅 API 文档，手动添加用户权限。如果需要添加多个用户权限，则按照上述步骤继续添加。各权限将并列显示在 AndroidManifest.xml 文件中。常用的用户权限如表 7-4 所示。

表 7-4　常用用户权限

用户权限名称	用户权限作用
android.permission.CALL_PHONE	允许一个程序不通过用户拨号，直接初始化一个电话拨号。但运行时，会通过用户界面的方式要求用户确认
android.permission.READ_CONTACTS	允许程序读取用户联系人数据
android.perimission.READ_PHONE_STATE	允许程序读取电话状态
android.permission.READ_SMS	允许程序读取短信息
android.permission.RECORD_AUDIO	允许程序录制音频
android.permission.SEND_SMS	允许程序发送 SMS 短信
android.permission.VIBRATE	允许访问振动设备
android.permission.WRITE_CONTACTS	允许程序写入但不读取用户的联系人数据
android.permission.WRITE_OWNER_DATA	允许一个程序写入但不读取所有者数据
android.permission.WRITE_SMS	允许程序写短信
android.permission.WRITE_SYNC_SETTINGS	允许程序写入同步设置
android.permission.WRITE_GSERVICES	允许程序修改 Google 服务地图
android.permission.WRITE_SETTINGS	允许程序读取或写入系统设置
android.permission.RECEIVE_SMS	允许程序监控接收短信息，并进行记录或处理
android.permission.READ_OWNER_DATA	允许程序读取所有者数据

运行程序，单击“获取信息”按钮，触发单击事件。相关信息输出在 LogCat 面板中，如图 7.16 所示。

```
com.example.telephony...   System.out    phoneType1
com.example.telephony...   System.out    phoneNumber15555215554
com.example.telephony...   System.out    deviceId000000000000000
com.example.telephony...   System.out    simNumber89014103211118510720
com.example.telephony...   System.out    netWorkCountryAndroid
```

图 7.16　LogCat 面板中输出的信息

7.4.2　短信管理器 SmsManager

SmsManager 是 Android 提供的与短信相关的系统服务。该 Service 使用流程介绍如下。

1. 获取 SmsManager 系统服务

使用该 Service 之前，首先需要获取 SmsManager 系统服务。语法如下：

```
smsManager = SmsManager.getDefault();
```

2. 调用 SmsManager 相关方法

SmsManager 类中，提供了与短信相关的方法，如表 7-5 所示。

3. 声明 PendingIntent 对象

在以上两个方法中，都使用到了 PendingIntent 参数。因此在调用 SmsManager 相关方法之前，需要声明一个 PendingIntent 对象。

表 7-5 SmsManager 相关方法

方 法 名 称	方 法 说 明
public ArrayList<String> divideMessage(String text)	当短信超过 SMS 消息的最大长度时,将短信分割为几块
public void sendDataMessage(String destinationAddress, String scAddress, short destinationPort, byte[] data, PendingIntent sentIntent, PendingIntent deliveryIntent)	通过短信,发送数据到一个特定的应用程序端
public void sendTextMessage(String destinationAddress, String scAddress, String text, PendingIntent sentIntent, PendingIntent deliveryIntent):	发送一个文本短信
public void sendMultipartTextMessage(String destinationAddress, String scAddress, ArrayList<String> parts, ArrayList<PendingIntent> sentIntents, ArrayList<PendingIntent> deliveryIntents)	发送多条文字短信

PendingIntent 这个类用于处理即将发生的事情。要得到一个 pendingIntent 对象,需要使用方法类的静态方法 getActivity(Context, int, Intent, int)、getBroadcast(Context, int, Intent, int)、getService(Context, int, Intent, int),它们分别对应着 Intent 的 3 个行为:跳转到一个 Activity 组件、打开一个广播组件、打开一个服务组件。

```
PendingIntent pIntent = PendingIntent.getActivity
                                            // 获得 PendingIntent 实例对象
                       (context,             // 当前上下文环境
                        requestCode,         // 请求码
                        intent,              // 启动意图
                        flags);              // 意图标志
```

【示例 7-6】 使用短信系统服务。新建项目 SmsManager,在程序中,添加两个编辑框,一个用于输入收信人号码,一个用于输入短信内容。再添加一个按钮,单击即可发送短信。逻辑代码如下:

```
public class SmsManagerActivity extends Activity {
    //SmsManager 对像
    private SmsManager smsManager;
    //收信人号码和短息内容编辑框
    private EditText number,content;
    //发送按钮
    private Button send;
    @Override
    public void onCreate(Bundle savedInstanceState) {
        super.onCreate(savedInstanceState);
        setContentView(R.layout.activity_sms_manager);
        //获取 SmsManger 系统服务
        smsManager = SmsManager.getDefault();
        number = (EditText)findViewById(R.id.et1);
        content = (EditText)findViewById(R.id.et2);
        send = (Button)findViewById(R.id.button1);
        send.setOnClickListener(new OnClickListener() {
            public void onClick(View v) {
                // TODO Auto-generated method stub
                //声明 PendingIntent 对象
```

```
            PendingIntent pIntent = PendingIntent.getActivity
        (SmsManagerActivity.this, 0, new Intent(), 0);
            //发送短信
            smsManager.sendTextMessage(number.getText().toString(),
                null,
                content.getText().toString(),
                pIntent,
                null);
            Toast.makeText(SmsManagerActivity.this, "已发送", Toast.
            LENGTH_LONG).show();
        }
    });
}
```

在 AndroidManifest.xml 中添加 android.permission.SEND_SMS 用户权限。运行程序，效果如图 7.17 所示。

图 7.17　发送短信息

7.4.3　音频管理器 AudioManager

AudioManager 是用来控制手机铃声和音量的系统服务。该 Service 使用流程介绍如下。

1. 获取 AUDIO_SERVICE 服务

使用该 Service 之前，首先需要获取 AUDIO_SERVICE 服务。语法如下：

```
AudioManager audioManager;                    // 声明 AudioManager 对象
audioManager = (AudioManager)getSystemService(Context.AUDIO_SERVICE);
                                              // 获得 AUDIO_SERVICE 系统服务
```

因为 getSystemService(String name)方法返回值类型是 Object，所以需要强制转换类型为 AudioManager 类。

2. 调用 AudioManager 相关方法

获取了 AudioManager 实例对象后，我们就可以调用它的相关方法。AudioManager 提供了一系列控制手机音量的方法，如表 7-6 所示。

第 7 章 Service 与 BroadcastReceiver

表 7-6　AudioManager 常用相关方法

方 法 名 称	方 法 说 明
adjustStreamVolume(int streamType, int direction, int flags)	调整手机指定类型的声音
setMode(int mode)	设置声音模式
setRingerMode(int ringerMode)	设置铃声模式
setStreamMute(int streamType, boolean state)	设置静音模式
setStreamVolume(int streamType, int index, int flags)	设定手机指定类型的音量值 streamType：指定声音类型 direction：控制声音的大小 flags：调整声音的标志

【示例 7-7】 使用 AUDIO_SERVICE 服务。新建项目 AudioManager，获取 AUDIO_SERVICE 服务。在程序中添加 3 个按钮：Play、Up、Down。在 Play 按钮的监听事件中调用 MediaPlayer 类播放音乐；在 Up 按钮的监听事件中使用 AUDIO_SERVICE 服务增大音量；在 Down 按钮的监听事件中使用 AUDIO_SERVICE 服务减小音量；通过喇叭形状的开关按钮设置静音。

在项目的 res 目录下新建文件夹 raw，添加音频文件到 raw 中供程序使用。如图 7.18 所示。

注意：音频文件的名称由 a~z，0~9 之间的字符组成，命名时要遵守命名规范，否则程序会报错。

图 7.18　添加音频文件

逻辑代码如下：

```
public class AudioManagerActivity extends Activity {
    private Button play,up,down;
    private ToggleButton off;
    private AudioManager audioManager ;
    @Override
    public void onCreate(Bundle savedInstanceState) {
        super.onCreate(savedInstanceState);
        setContentView(R.layout.activity_audio_manager);
        //获取系统服务
        audioManager = (AudioManager)getSystemService(Context.AUDIO_SERVICE);
        play =(Button)findViewById(R.id.button1);
        play.setOnClickListener(new OnClickListener() {
            public void onClick(View v) {
                // TODO Auto-generated method stub
                //根据给定资源创建 MediaPlayer 对象
                MediaPlayer mediaPlayer = MediaPlayer.create
                    (AudioManagerActivity.this, R.raw.goodlife);
                //播放音乐
                mediaPlayer.start();
            }
        });
        up = (Button)findViewById(R.id.button2);
        up.setOnClickListener(new OnClickListener() {
            public void onClick(View v) {
                // TODO Auto-generated method stub
                audioManager.setStreamVolume(
```

· 155 ·

```
                    //指定为手机音乐类型
                    AudioManager.STREAM_MUSIC,
                    //增大音量
                    AudioManager.ADJUST_RAISE,
                    //音量变化显示进度条
                    AudioManager.FLAG_SHOW_UI);
            }
        });
        down = (Button)findViewById(R.id.button3);
        down.setOnClickListener(new OnClickListener() {
            public void onClick(View v) {
                // TODO Auto-generated method stub
                audioManager.setStreamVolume(
                    //指定为手机音乐类型
                    AudioManager.STREAM_MUSIC,
                    //减小音量
                    AudioManager.ADJUST_LOWER,
                    //音量变化显示进度条
                    AudioManager.FLAG_SHOW_UI);
            }
        });
        off = (ToggleButton)findViewById(R.id.toggleButton1);
        off.setOnCheckedChangeListener(new OnCheckedChangeListener() {

            public void onCheckedChanged(CompoundButton buttonView,
            boolean isChecked) {
                // TODO Auto-generated method stub
                //设置为静音模式
                audioManager. setStreamMute(AudioManager.STREAM_MUSIC,
                isChecked);
            }
        });
    }
}
```

运行程序，单击 Play 按钮播放音乐；单击 Up 按钮出现进度条，可以拖动增大音量；单击 Down 按钮出现进度条，可以拖动减小音量，单击 Off 按钮设置为静音模式。效果如图 7.19 所示。

图 7.19　AudioManager 效果图

7.4.4 振动器 Vibrator

Android 手机中的振动由 Vibrator 实现。在与用户交互时，常常会用到振动功能，尤其在游戏中，应用更加广泛，比如爆炸、碰撞等。

1. 获取 VIBRATOR_SERVICE 系统服务

使用该 Service 之前，首先需要获取 VIBRATOR_SERVICE 系统服务。语法如下：

```
Vibrator vibrator;                                           // 声明Vibrator对象
vibrator = (Vibrator)getSystemService(Context.VIBRATOR_SERVICE);
                                                             // 获得Vibrator系统服务
```

因为 getSystemService(String name)方法返回值类型是 Object，所以需要强制转换类型为 Vibrator 类。

2. 调用 Vibrator 相关方法

获取了 Vibrator 实例对象后，我们就可以调用它的相关方法。Vibrator 比较简单，只提供了少量的方法，如表 7-7 所示。

表 7-7 Vibrator 相关方法

方法名称	方法说明
void cancel()	关闭振动
boolean hasVibrator()	检测是否有振动硬件
void vibrate(long milliseconds)	在一定的时间内振动

【示例 7-8】 使用 VIBRATOR_SERVICE 系统服务。新建项目 Vibrator，在程序中添加两个按钮，一个使手机开始振动，一个使手机关闭震动。

逻辑代码如下：

```java
public class VibratorActivity extends Activity {
    private Vibrator vibrator;
    private Button start,stop;
    @Override
    public void onCreate(Bundle savedInstanceState) {
        super.onCreate(savedInstanceState);
        setContentView(R.layout.activity_vibrator);
        //获取系统服务
        vibrator = (Vibrator)getSystemService(Context.VIBRATOR_SERVICE);
        start = (Button)findViewById(R.id.button1);
        //start 按钮监听
        start.setOnClickListener(new OnClickListener() {
            public void onClick(View v) {
                // TODO Auto-generated method stub
                //振动3000毫秒
                vibrator.vibrate(3000);
                Toast.makeText(VibratorActivity.this, "手机振动", Toast.LENGTH_LONG).show();
            }
        });
            stop = (Button)findViewById(R.id.button2);
        //stop 按钮监听
        stop.setOnClickListener(new OnClickListener() {
```

```
    public void onClick(View v) {
        // TODO Auto-generated method stub
        //关闭振动
        vibrator.cancel();
        Toast.makeText(VibratorActivity.this, "振动已关闭",
        Toast.LENGTH_LONG).show();
    }
});
  }
}
```

在 AndroidManifest.xml 中添加 android.permission.VIBRATE 用户权限。运行程序，单击 Start 按钮，手机开始振动，Toast 显示提示消息。单击 Stop 按钮，手机停止振动，Toast 显示提示消息。效果如图 7.20 所示。

图 7.20　Vibrator 效果图

注意：Vibrator 在模拟器上没有效果。程序完成后，请在真机上测验。

7.5　广播接收者 BroadcastReceiver

在 Android 中，广播 Broadcast 是一种广泛运用在应用程序之间的用于传送消息的机制。而 BroadcastReceiver 是用来过滤接收并响应 Broadcast 的一类组件。它可以监听系统全局的广播消息，非常方便地实现系统中不同组件之间通信。

7.5.1　开发 BroadcastReceiver

BroadcastReceiver 的运行机理非常简单。开发过程如下：
- 开发 BroadcastReceiver 的子类，重写其中的 onReceive()方法；
- 注册 BroadcastReceiver 对象；
- 将需要广播的消息封装到 Intent 中，然后调用方法发送出去；
- 通过 IntentFilter 对象过滤 Intent，处理与其匹配的广播。

1. 注册 BroadcastReceiver

与其他组件相同，BroadcastReceiver 在使用之前，也需要在 AndroidManifest.xml 中注

册——静态注册。静态注册的特点是，不管该应用程序是否处于活动状态，都会进行监听。注册 BroadcastReceiver 使用<receiver></receiver>标签。然后通过<intent-filter></intent-filter>标签设置过滤条件。<receiver></receiver>与其他组件（例如<activity></activity>）并列位于<application></application>标签内，为同一个应用程序所用。语法如下：

```xml
<receiver android:name=" ">                <!--广播接收者名称-->
    <intent-filter >
        <action android:name=" " />        <!-- Intent-filter 过滤条件-->
    </intent-filter>
</receiver>
```

BroadcastReceiver 也可以在代码中注册——动态注册。首先创建 IntentFilter 对象，并设置过滤条件，然后通过 Context.registerReceiver()方法来注册，通过 Context.unregisterReceiver()方法取消注册。动态注册的特点是，在代码中进行注册后，当应用程序关闭，就不再进行监听。

```
IntentFilter intentFilter = new IntentFilter (String action);
registerReceiver (BroadcastReceiver receiver, IntentFilter,filter );
```

2. 发送广播

Android 系统提供了 Context.sendBroadcast()和 Context.sendOrderedBroadcast()两种方法发送广播或有序广播，供 BroadcastReceiver 接收并处理。

- Context.sendBroadcast()发送的广播，所有满足条件的 BroadcastReceiver 都会执行其 onReceive()方法来处理响应。
- Context.sendOrderedBroadcast 发送的有序广播。会根据 BroadcastReceiver 注册时 IntentFilter 的优先级顺序来执行 onReceive()方法。优先级在<intent-filter>的 android:priority 中声明，也可以在代码中通过 IntentFilter.setPriority()方法设置。数越大优先级别越高。

【示例7-9】 调用 Context.sendBroadcast()发送广播。新建项目 BroadcastReceiver1，在界面添加按钮发送广播。创建 BroadcastReceiverDemo 类继承于 BroadcastReceiver，在 BroadcastReceiver1Activity 中动态注册 BroadcastReceiverDemo 接收广播。

BroadcastReceiverDemo：

```java
//创建 BroadcastReceiverDem 类继承于 BroadcastReceiver
public class BroadcastReceiverDemo extends BroadcastReceiver {
    @Override
    //重写 onReceive()方法
    public void onReceive(Context context, Intent intent) {
        // TODO Auto-generated method stub
        //接收广播
        String bundle = intent.getStringExtra("broadcast");
        System.out.println("接收到： " + bundle);
    }
}
```

BroadcastReceiver1Activity：

```java
public class BroadcastReceiver1Activity extends Activity {
    private Button button;
```

```java
    private Intent intent;
    @Override
    public void onCreate(Bundle savedInstanceState) {
        super.onCreate(savedInstanceState);
        setContentView(R.layout.activity_broadcast_receiver1);
        //过滤条件
        IntentFilter intentFilter = new IntentFilter("BROADCAST_RECEIVER_DEMO");
        //注册广播
        registerReceiver(new BroadcastReceiverDemo(), intentFilter);
        intent = new Intent();
        //与过滤条件匹配
        intent.setAction("BROADCAST_RECEIVER_DEMO");
        Bundle bundle = new Bundle();
        //广播内容
        bundle.putString("broadcast", "DEMO");
        intent.putExtras(bundle);
        button = (Button)findViewById(R.id.button1);
        button.setOnClickListener(new OnClickListener() {
            public void onClick(View v) {
                // TODO Auto-generated method stub
                //发送广播
                sendBroadcast(intent);
            }
        });
    }
}
```

运行程序，单击"发送广播"按钮，输出广播内容在 LogCat 面板上。如图 7.21 所示。

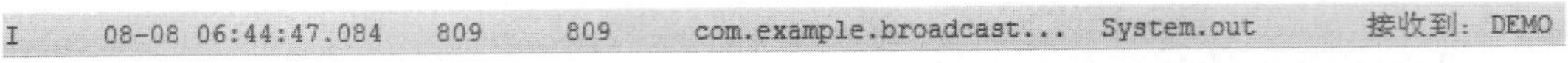

```
I   08-08 06:44:47.084    809      809    com.example.broadcast... System.out    接收到: DEMO
```

图 7.21　LogCat 面板上输出的信息

【示例 7-10】　调用 Context.sendOrderedBroadcast 发送有序广播。新建项目 BroadcastReceiver2，在界面添加按钮发送有序广播。分别创建 BroadcastReceiverDemoOne 和 BroadcastReceiverDemoTwo 继承于 BroadcastReceiver，在 AndroidManifest.xml 中静态注册 BroadcastReceiverDemoOne 和 BroadcastReceiverDemoTwo 接收广播。

BroadcastReceiverDemoOne：

```java
public class BroadcastReceiverDemoOne extends BroadcastReceiver {
    @Override
    public void onReceive(Context context, Intent intent) {
        // TODO Auto-generated method stub
        //接收到广播后打印 DemoOne
        System.out.println("DemoOne");
    }
}
```

BroadcastReceiverDemoTwo：

```java
public class BroadcastReceiverDemoTwo extends BroadcastReceiver {
    @Override
    public void onReceive(Context context, Intent intent) {
        // TODO Auto-generated method stub
        //接收到广播后打印 DemoTwo
        System.out.println("DemoTwo");
```

        }
}
```

BroadcastReceiver2Activity：

```java
public class BroadcastReceiver2Activity extends Activity {
    private Button button;
    @Override
    public void onCreate(Bundle savedInstanceState) {
        super.onCreate(savedInstanceState);
        setContentView(R.layout.activity_broadcast_receiver2);
        button = (Button)findViewById(R.id.button1);
        button.setOnClickListener(new OnClickListener() {
            public void onClick(View v) {
                // TODO Auto-generated method stub
                Intent intent = new Intent();
                //过滤条件
                intent.setAction("BROADCAST_RECEIVER_DEMO");
                //发送有序广播
                sendOrderedBroadcast(intent, null);
            }
        });
    }
}
```

AndroidManifest.xml：

```xml
<manifest xmlns:android="http://schemas.android.com/apk/res/android"
    package="com.example.broadcastreceiver2"
    android:versionCode="1"
    android:versionName="1.0" >

    <uses-sdk
        android:minSdkVersion="8"
        android:targetSdkVersion="15" />

    <application
        android:icon="@drawable/ic_launcher"
        android:label="@string/app_name"
        android:theme="@style/AppTheme" >
        <activity
            android:name=".BroadcastReceiver2Activity"
            android:label="@string/title_activity_broadcast_receiver2" >
            <intent-filter>
                <action android:name="android.intent.action.MAIN" />

                <category android:name="android.intent.category.LAUNCHER" />
            </intent-filter>
        </activity>
        <!--注册 BroadcastReceiverDemoOne -->
        <receiver android:name="BroadcastReceiverDemoOne">
        <!--设置优先级为1-->
            <intent-filter android:priority="1">
                <!--匹配成功-->
                <action android:name="BROADCAST_RECEIVER_DEMO" />
            </intent-filter>
        </receiver>
        <!--注册 BroadcastReceiverDemoTwo -->
        <receiver android:name="BroadcastReceiverDemoTwo">
```
```

```xml
            <!--设置优先级为10-->
            <intent-filter android:priority="10">
                <!--匹配成功-->
                <action android:name="BROADCAST_RECEIVER_DEMO" />
            </intent-filter>
        </receiver>
    </application>
</manifest>
```

运行程序，单击"发送广播"按钮，查看 LogCat 面板中的输出结果。如图 7.22 所示。

```
I    08-08 07:48:10.014    1186    1186    com.example.broadcast...    System.out    DemoTwo
I    08-08 07:48:10.047    1186    1186    com.example.broadcast...    System.out    DemoOne
```

图 7.22 LogCat 面板输出的结果

由上可知，BroadcastReceiverDemoTwo 优先级较高，则优先执行。BroadcastReceiverDemoOne 优先级较低，则稍后执行。

7.5.2 接收系统广播信息

除了接收用户发送的广播之外，BroadcastReceiver 还有一个重要的功能是接收系统广播。如果应用需要在系统特定时刻执行某些操作，就可以通过监听系统广播来实现。Android 常见的广播 Action 常量如表 7-8 所示。

表 7-8 Android 常见的广播 Action 常量

常 量 名 称	常 量 值	说 明
android.intent.action.BOOT_COMPLETED	ACTION_BOOT_COMPLETED	系统启动
android.intent.action.ACTION_TIME_CHANGED	ACTION_TIME_SET	时间改变
android.intent.action.ACTION_DATE_CHANGED	ACTION_DATE_CHANGED	日期改变
android.intent.action.ACTION_TIMEZONE_CHANGED	ACTION_TIMEZONE_CHANGED	时区改变
android.intent.action.ACTION_BATTERY_LOW	ACTION_BATTERY_LOW	电量低
android.intent.action.ACTION_MEDIA_EJECT	ACTION_MEDIA_EJECT	插入或拔出外部媒体
android.intent.action.ACTION_MEDIA_BUTTON	ACTION_MEDIA_BUTTON	按下多媒体键
android.intent.action.ACTION_PACKAGE_ADDED	ACTION_PACKAGE_ADDED	添加包
android.intent.action.ACTION_PACKAGE_REMOVED	ACTION_PACKAGE_REMOVED	删除包

【示例 7-11】 接收系统广播，监控系统日期的变化。当用户更改系统日期，该程序会自动启动，提示用户日期被更改。

MyReceiver：

```java
public class MyReceiver extends BroadcastReceiver {
    @Override
    public void onReceive(Context context, Intent intent) {
        // TODO Auto-generated method stub
        //创建 Intent 对象
        Intent i = new Intent(context, TimeActivity.class);
        //设置 Intent 的 flag
```

```
        i.setFlags(Intent.FLAG_ACTIVITY_NEW_TASK);
        //启动Activity
        context.startActivity(i);
    }
}
```

TimeActivity:

```
public class TimeActivity extends Activity {
    private Button button;
    @Override
    public void onCreate(Bundle savedInstanceState) {
        super.onCreate(savedInstanceState);
        setContentView(R.layout.activity_time);
        button = (Button)findViewById(R.id.button1);
        button.setOnClickListener(new OnClickListener() {
            public void onClick(View v) {
                // TODO Auto-generated method stub
                // 退出程序
                System.exit(0);
            }
        });
    }
}
```

AndroidManifest.xml:

```
<manifest xmlns:android="http://schemas.android.com/apk/res/android"
    package="com.example.time"
    android:versionCode="1"
    android:versionName="1.0" >

    <uses-sdk
        android:minSdkVersion="8"
        android:targetSdkVersion="15" />
    <uses-permission android:name="android.permission.WRITE_SETTINGS"/>

    <application
        android:icon="@drawable/ic_launcher"
        android:label="@string/app_name"
        android:theme="@style/AppTheme" >
        <activity
            android:name=".TimeActivity"
            android:label="@string/title_activity_time" >
            <intent-filter>
                <action android:name="android.intent.action.MAIN" />
                <category android:name="android.intent.category.LAUNCHER" />
            </intent-filter>
        </activity>
        <!--注册MyReceiver -->
        <receiver android:name="MyReceiver">
            <intent-filter>
                <!--日期更改-->
                <action android:name="android.intent.action.DATE_CHANGED" />
                <category android:name="android.intent.category.HOME" />
            </intent-filter>
        </receiver>
    </application>

</manifest>
```

先运行一次程序,再将其关闭(相当于将应用程序安装到模拟器中)。然后,在模拟器的设置中修改日期,此时会自动启动提示程序,如图 7.23 所示。

图 7.23　修改系统日期

7.6　小　　结

通过本章的学习,读者应该能够很好地掌握 Android 的后台服务机制和消息广播机制。本章的重点是 Service 的开发与通信,以及 BroadcastReceiver 的开发。系统提供的 Service 功能强大,如何结合 BroadcastReceiver 一起使用,是本章学习的难点。二者的运用都涉及添加对应的用户权限,也需要读者认真掌握。

7.7　习　　题

新建项目 Time,监控手机的系统时间的变化。当系统时间被修改时,通过广播提示,如图 7.24 所示。

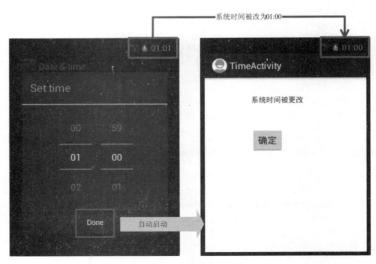

图 7.24　监控系统时间

【分析】本题目考查了读者对广播的掌握。首先在 AndroidManifest.xml 文件中添加用户权限"android.intent.action.ACTION_TIME_CHANGED"，允许程序修改系统时间。然后创建 MyReceiver 继承于 BroadcastReceiver，重写 onReceiver()方法。在该方法中创建 Intent 对象，并设置其 Flag 值。最后启动相应 Activity。

【核心代码】本题的核心代码如下所示。

```java
public class MyReceiver extends BroadcastReceiver {
    @Override
    public void onReceive(Context context, Intent intent) {
        // TODO Auto-generated method stub
        Intent i = new Intent(context, TimeActivity.class);
        i.setFlags(Intent.FLAG_ACTIVITY_NEW_TASK);
        context.startActivity(i);
    }
}
```

第 8 章 Android 数据存储

数据存储是应用程序最基本的问题，任何企业系统、应用软件都必须解决这一问题，数据存储必须以某种方式保存，不能丢失，并且能够有效、简单地使用和更新这些数据。Android 提供了以下几种方式供开发者存取数据：SharedPreferences 存储、文件存储、SQLite 数据库存储、ContentProvider。本章将详细讲解这几种存储方式。

8.1 Android 中存储概要

在 Android 中一共提供了 5 种数据存储方式，它们各有各的特点。
- SharedPreferences：它是一个较轻量级的存储数据的方法。用来存储 key-value 形式的数据，只可以用来存储基本的数据类型。
- File：文件存储方式是一种比较常见的存储方式，是 Android 中读取/写入文件的方法，和 Java 中实现程序的 I/O 一样，它提供了 FileInputStream 和 FileOutputStream 方法来对文件进行操作。
- SQLite：Android 提供的一个标准数据库，并支持 SQL 语句。
- Network：主要通过网络来存储和获取数据。由于使用较少，本书不做介绍。
- ContentProvider：数据共享，它是应用程序之间唯一共享数据的一个方法。一个程序可以通过数据共享来访问另一个程序的数据。

了解了 Android 中数据的数据存储形式之后，我们就可以根据程序的需要来选择最合适的存储方式了。本节将对以上几种存储方式进行详细的介绍。

8.2 键值对存储 SharedPreferences

SharedPreferences 是 Android 平台上一个轻量级的存储类。它用来存储一些简单的 Key-Value 名值对。它的 value 值只能是 int、long、boolean、String 和 float 类型。在应用程序中主要保存一些常用的配置信息。

8.2.1 SharedPreferences 是什么

SharedPreferences 是一个接口，语法如下：

```
SharedPreferences spf = getSharedPreferences(String name,int mode);
```

name 表示存储数据的文件名；mode 表示对数据操作的几种方式，它的可选值有以下几个：
- MODE_PRIVATE，指定该 SharedPreferences 里的数据只能被本应用程序读写；

- MODE_WORLD_READABLE，指定该 SharedPreferences 里的数据可以被其他应用程序读，但不能写；
- MODE_WORLD_WRITEABLE，指定该 SharedPreferences 里的数据可以被其他应用程序读写。

SharedPreferences 提供了一系列方法来获取应用程序中的数据，如表 8-1 所示。

表 8-1　SharedPreferences 相关方法

方 法 名 称	方 法 说 明
boolean contains(Stirng key)	判断 SharedPreferences 是否包含特定 key 的数据
abstract Map<String,?> getAll()	获取 SharedPreferences 里的全部 Key-Value 对
boolean getBoolean(String key, boolean defValue)	获取 SharedPreferences 里指定 key 对应的 boolean 值
int getInt(String key, int defValue)	获取 SharedPreferences 里指定 key 对应的 int 值
float getFloat(String key, float defValue)	获取 SharedPreferences 里指定 key 对应的 float 值
long getLong(String key, long defValue)	获取 SharedPreferences 里指定 key 对应的 long 值
String getString(String key, string defValue)	获取 SharedPreferences 里指定 key 对应的 String 值

SharedPreferences 对象本身只能获取数据，并不支持数据的存储和修改。存储和修改是通过 SharedPreferences.Editor 对象实现。获取 Editor 实例对象，需要调用 SharedPreferences.Editor edit()方法。SharedPreferences.Editor 相关方法如表 8-2 所列。

表 8-2　SharedPreferences.Editor 相关方法

方 法 名 称	方 法 说 明
SharedPreferences.Editor edit()	创建一个 Editor 对象
SharedPreferences.Editor clear()	清空 SharedPreferences 里所有数据
SharedPreferences.Editor putString(String key,String value)	向 SharedPreferences 存入指定 key 对应的 String 值
SharedPreferences.Editor putInt(String key, int value)	向 SharedPreferences 存入指定 key 对应的 Int 值
SharedPreferences.Editor putFloat(String key, float value)	向 SharedPreferences 存入指定 key 对应的 String 值
SharedPreferences.Editor putLong(String key, long value)	向 SharedPreferences 存入指定 key 对应的 String 值
SharedPreferences.Editor putBoolean(String key, boolean value)	向 SharedPreferences 存入指定 key 对应的 boolaen 值
SharedPreferences.Editor remove(String key)	删除 SharedPreferences 指定 key 对应的数据
boolean commit()	当 Editor 编辑完之后，调用该方法提交

8.2.2　SharedPreferences 实现数据存储

使用 SharedPreferences 存储数据的步骤如下：
- 调用 context.getSharedPreferences(String name, int mode)方法获取 SharedPreferences 对象；
- 利用 SharedPreferences.Editor edit()方法获取 Editor 对象；
- 通过 Editor 对象存储 key-value 名值对数据；
- 通过 commit()方法提交数据。

【示例 8-1】使用 SharedPreferences 存储数据，数据保存在 data/data/项目名/shared_prefs 路径下。

逻辑代码如下：

```java
public class SharedPreferencesActivity extends Activity {
    @Override
    public void onCreate(Bundle savedInstanceState) {
        super.onCreate(savedInstanceState);
        setContentView(R.layout.activity_shared_preferences);
        //获取 SharedPreferences 对象
        SharedPreferences spf = getSharedPreferences(
        //保存数据的文件
        "save",
        //指定该 SharedPreferences 的数据可被其他程序读取
        MODE_WORLD_READABLE);
        //创建 Editor 对象
        SharedPreferences.Editor edit = spf.edit();
        //保存数据到 SharedPreferences
        edit.putString("abc", "SharedPreferencesc 存储数据");
        //提交数据
        edit.commit();
    }
}
```

运行程序,查看 save 文件中数据存储情况。在 Eclipse 中,单击右上角的 Open Perspective 按钮,打开 DDMS 视图。在 File Explorer 面板中,进入 data/data/com.example.SharedPreferences/shared_prefs 路径找到 save.xml 文件,导出并查看数据信息。详细步骤如图 8.1 所示。

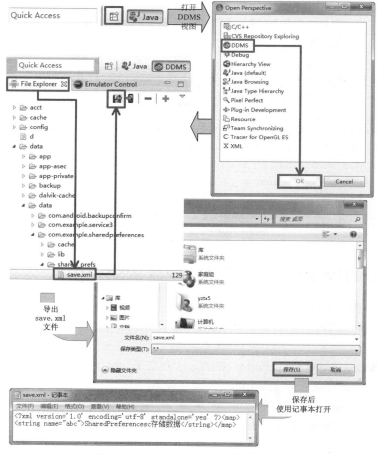

图 8.1 SharedPreferences 存储数据

8.3 File 存储

Android 也支持使用文件存取数据。File 的存储和读取主要使用的是 Java 中的 I/O 流，通过输出流存储数据，通过输入流读取数据。

8.3.1 File 实现数据读取

Context 提供了下面两个方法打开应用程序中数据文件夹里的文件 I/O 流。
- FileInputStream openFileInput(String name)：打开应用程序中数据文件夹里 name 文件对应的输入流。语法如下：

```
FileInputStream fis = openFileInput(String name);
```

- FileOutputStream openFileOutput(String name int mode)：打开应用程序中数据文件夹里 name 文件对应的输出流。语法如下：

```
FileOutputStream fos = openFileOutput(String name, int mode);
```

第二个方法中的 mode 参数支持以下几个值：
- MODE_PRIVATE，该文件只能被当前程序读写；
- MODE_APPEND，该文件内容可被追加；
- MODE_WORLD_READABLE，该文件内容可被其他程序读取；
- MODE_WORLD_WRITEABLE，该文件内容可被其他程序读写。

【示例 8-2】 读写文件中的数据。新建一个项目 FileStore，在界面添加一个编辑框，用于输入数据信息。然后添加两个按钮，一个用于保存数据，另一个用于读取数据。再添加一个文本框设置为空，用于显示读取出来的数据信息。

逻辑代码如下：

```
public class FileStoreActivity extends Activity {
    //保存按钮
    private Button button1;
    //获取按钮
    private Button button2;
    //输入信息
    private EditText edt;
    //获取信息显示
    private TextView tv;
    //输入流
    private FileInputStream fis;
    //输出流
    private FileOutputStream fos;
    //文件名
    private String FILE_NAME="file";
    @Override
    public void onCreate(Bundle savedInstanceState) {
        super.onCreate(savedInstanceState);
        setContentView(R.layout.activity_file_store);

        button1 = (Button)findViewById(R.id.button1);
        button2 = (Button)findViewById(R.id.button2);
```

```
        edt = (EditText)findViewById(R.id.edt);
        tv = (TextView)findViewById(R.id.tv);
        try {
         //以追加方式打开输出流
            fos = openFileOutput(FILE_NAME, Context.MODE_APPEND);
            //打开输入流
            fis = openFileInput(FILE_NAME);
        } catch (FileNotFoundException e) {
            e.printStackTrace();
            Toast.makeText(this, "文件不存在", Toast.LENGTH_SHORT).show();
        }
    }
    public void OnClickMethod(View v){
     if (button1.getId()==v.getId()) {
        //获取编辑框数据
        String msgsave=edt.getText().toString();
        try {
                //将数据写入文件
                fos.write(msgsave.getBytes());
                Toast.makeText(this, "保存成功", Toast.LENGTH_SHORT).show();
                fos.close();
            } catch (IOException e) {
                e.printStackTrace();
            }
    }else if (button2.getId()==v.getId()) {
        //声明一个长度为 200 的字节数组
        byte[] b=new byte[200];
        try {
                //读取数据到 b 中
                fis.read(b);
                //将数据显示在 TextView 中
                tv.setText(new String(b));
                fis.close();
            } catch (IOException e) {
                // TODO Auto-generated catch block
                e.printStackTrace();
            }
        }
    }
}
```

运行程序。在编辑框内输入信息，单击"保存"按钮保存信息，保存成功则 Toast 提示"保存成功"；然后单击"读取"按钮，获取信息，并显示在文本框中，如图 8.2 所示。

图 8.2　存储数据到文件

在编辑框重新输入内容,单击"保存"按钮保存信息。然后,单击"读取"按钮,显示两次输入的内容在文本框中,表示追加信息到文件保存成功,如图 8.3 所示。

使用文件存储的数据保存在 data/data/项目名/files 路径下,同样可以导出文件查看数据存储状况,如图 8.4 所示。

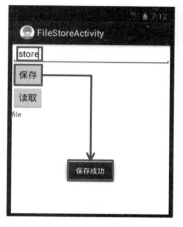

图 8.3　追加数据到文件　　　　　　图 8.4　文件存储的数据

8.3.2　File 实现 SD 卡中数据的读写

由于手机内存有限,有时并不能满足用户的需求。为了更好地存、取应用程序的大文件数据,应用程序可以读写 SD 卡中的数据。SD 卡大大扩充了手机的存储能力。读写 SD 卡中数据的步骤介绍如下。

(1)调用 Environment.getExternalStorageStata()方法,判断手机是否插入了 SD 卡,并且应用程序是否具有读写 SD 卡数据的权限。如果手机插入了 SD 卡,并且应用程序具有读写 SD 卡数据的权限,那么下面的程序将返回 true。

```
Environment.getExternalStorageState().equals(Environment.MEDIA_MOUNTED)
```

(2)调用 Environment.getExternalStorageDirectory()方法,获取 Android 外部存储器,即 SD 卡目录。

```
File sdFile = Environment.getExternalStorageDirectory();
```

(3)调用 FileInputStream、FileOutputStream、FileReader 或 FileWriter 读写 SD 卡里的文件。

在读写 SD 卡中的文件时,需要注意以下几点。

第一,手机要插有 SD 卡。对于模拟器来说,可以创建虚拟设备的 SD 卡。在 Eclipse 的左上角,单击 Opens the Android Virtual Device Manager 按钮,打开 Android 虚拟设备管理器。选中项目所用的 AVD(如果没有 AVD 可选,则单击 Android Virtual Device Manager 视图中的 new 按钮新建 AVD,并设置 SD 卡),设置其 SD card 容量大小。然后,刷新并启动该模拟器。步骤如图 8.5 所示。

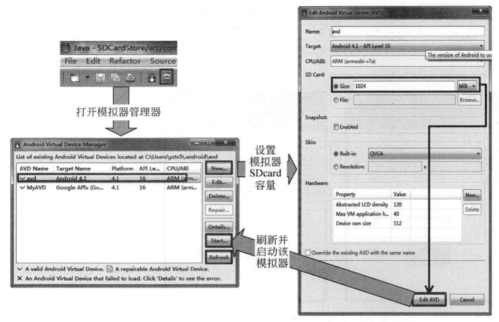

图 8.5 设置 AVD 的 SDcard 容量

第二，在 AndroidManifest.xml 文件中，添加用户权限。

- android.permission.MOUNT_UNMOUNT_FILESYSTEMS：允许应用程序在 SD 卡中创建或删除文件。
- android.permission.WRITE_EXTERNAL_STORAGE：允许向 SD 卡中写入数据。

第三，AVD 的虚拟 SD 卡中的文件，存储在 mnt/sdcard 目录下。查看时，需要打开 DDMS 视图的 File Explorer 面板，找到文件后导出查看。

【示例 8-3】 读写 SD 卡中的数据。新建一个项目 SDCardStore，在界面添加一个编辑框，用于输入信息。添加两个按钮，一个用于写入数据，另一个用于读取数据。

逻辑代码如下：

```
public class SDCardStoreActivity extends Activity {
    private EditText editText;
    private Button btnRead,btnWrite;
    @Override
    public void onCreate(Bundle savedInstanceState) {
        super.onCreate(savedInstanceState);
        setContentView(R.layout.activity_sdcard_store);
        editText = (EditText)findViewById(R.id.editText1);
        btnWrite = (Button)findViewById(R.id.button1);
        btnRead = (Button)findViewById(R.id.button2);
        btnWrite.setOnClickListener(new OnClickListener() {
            public void onClick(View v) {
                //判断是否插入 SD 卡
                String state = Environment.getExternalStorageState();
                //插入 SD 卡并且具有读写权限
                if (state.equals(Environment.MEDIA_MOUNTED)) {
                    //获取 SD 卡目录
                    File sdFile = Environment.getExternalStorageDirectory();
```

```
            //获取文件绝对路径
            String filePath = sdFile.getAbsolutePath();
            //获取编辑框数据
            String msg = editText.getText().toString();
            try {
                //将数据写入到SD卡中
                FileWriter fileWriter = new FileWriter(filePath+
                "/note.txt",true);
                BufferedWriter bWriter = new BufferedWriter
                (fileWriter);
                bWriter.write(msg);
                bWriter.flush();
                Toast.makeText(SDCardStoreActivity.this, "写入成功",
                        Toast.LENGTH_SHORT).show();
                bWriter.close();
                fileWriter.close();
            } catch (IOException e) {
                // TODO Auto-generated catch block
                e.printStackTrace();
            }
        }
    }
});
btnRead.setOnClickListener(new OnClickListener() {
    public void onClick(View v) {
        //判断是否插入SD卡
        String state = Environment.getExternalStorageState();
        //插入SD卡并且具有读写权限
        if (state.equals(Environment.MEDIA_MOUNTED)) {
            //获取SD卡目录
            File sdFile = Environment.getExternalStorageDirectory();
            //获取文件绝对路径
            String filePath = sdFile.getAbsolutePath();
            try {
                //读取SD卡数据，使用Toast显示
                FileReader fileReader = new FileReader(filePath+
                "/note.txt");
                BufferedReader bReader = new BufferedReader
                 (fileReader);
                String line;
                while ((line=bReader.readLine())!=null) {

                    Toast.makeText(SDCardStoreActivity.this, "内容
                    是:" + line,
                            Toast.LENGTH_SHORT).show();
                }
            } catch (IOException e) {
                // TODO Auto-generated catch block
                e.printStackTrace();
            }
        }
    }
});
    }
}
```

AndroidManifest.xml：

```
<manifest xmlns:android="http://schemas.android.com/apk/res/android"
```

```xml
        package="com.example.sdcardstore"
        android:versionCode="1"
        android:versionName="1.0" >
        <uses-sdk
            android:minSdkVersion="8"
            android:targetSdkVersion="15" />

        <!--添加用户权限-->
        <uses-permission android:name="android.permission.MOUNT_UNMOUNT_
FILESYSTEMS"/>
        <uses-permission android:name="android.permission.WRITE_EXTERNAL_
STORAGE"/>

        <application
            android:icon="@drawable/ic_launcher"
            android:label="@string/app_name"
            android:theme="@style/AppTheme" >
            <activity
                android:name=".SDCardStoreActivity"
                android:label="@string/title_activity_sdcard_store" >
                <intent-filter>
                    <action android:name="android.intent.action.MAIN" />
                    <category android:name="android.intent.category.LAUNCHER" />
                </intent-filter>
            </activity>
        </application>
</manifest>
```

运行程序。在编辑框内输入信息，单击 write 按钮保存信息，保存成功则 Toast 提示；然后单击 read 按钮，Toast 显示存储信息，如图 8.6 所示。

打开 DDMS 视图的 File Explorer 面板，进入 mnt/sdcard 路径找到 note.txt 文件，导出查看。如图 8.7 所示。

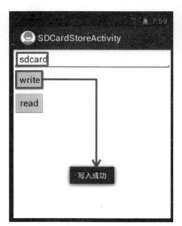

图 8.6　读写 SD 卡中的数据　　　　图 8.7　SD 卡中 note.txt 文件

8.4　SQLite 数据库存储

键值对存储和 File 存储都是适合于存储一些简单的、数据量较小的数据。如果要存储大量的数据，并且对其进行管理、升级、维护等，有可能还要随时添加、查看、删除和更新数据。这时，我们就需要采用 SQLite 数据库来进行数据存储了。

8.4.1 SQLite 数据库简介

SQLite 诞生于 2000 年 5 月，它是一款轻型数据库，它的设计目标是嵌入式的，而且目前已经在很多嵌入式产品中使用，它占用的资源非常少，在嵌入式设备中，可能只需要几百千字节的内存就足够了。也许这正是 Android 系统要采用 SQLite 数据的原因之一。SQLite 是用 C 语言编写的，它具有如下特征。

- 轻量级：SQLite 数据库和其他数据库不同，它不存在客户端和服务器端，使用它时只要能带上它的动态数据库就可以使用它的功能，且动态库也相当小。
- 跨平台：SQLite 目前支持大部分主流操作系统，它不仅能在计算机上运行，在手机操作系统上同样能够使用。
- 独立性：SQLite 数据库的引擎不需要依赖第三方软件，本身也不需要安装。
- 多语言接口：SQLite 数据库不止支持 Java 语言编程，还支持 C/C++、Python、.Net、Ruby、Perl 等，得到更多开发者的喜爱。
- 安全性：SQLite 数据库通过数据库级上的独占性和共享锁来实现独立事务处理。这意味着多个进程可以在同一时间从同一数据库读取数据，但只能有一个可以写入数据。

在 Android 平台下，可以通过 SQLiteDatabase 类的静态方法创建或打开数据库。主要有以下 3 种方法。

```
SQLiteDatabase database =                       // 声明一个 SQLiteDatabase 对象
    openDatabase(                               // 打开数据库
    String path,                                // path 数据库文件
    SQLiteDatabase.CursorFactory factory,
                                                // 用于生成一个游标对象,查询数据库时调用
    int flags                                   // 控制数据可访问模式
)
```

注意：当 flags 值置为 0 时，表示创建的数据库可读可写；当 flags 值置为 1 时，表示数据库只可读不可写。

```
SQLiteDatabase database =                       // 声明一个 SQLiteDatabase 对象
    openOrCreateDatabase(                       // 打开或创建（如果需要）数据库
    String path,                                // 数据库文件路径
    SQLiteDatabase.CursorFactory factory)
                                                // 用于生成一个游标对象,查询数据库时调用
SQLiteDatabase database =                       // 声明一个 SQLiteDatabase 对象
    openOrCreateDatabase(                       // 打开或创建（如果需要）数据库
    File file,                                  // File 数据库文件
    SQLiteDatabase.CursorFactory factory)
                                                // 用于生成一个游标对象,查询数据库时调用
```

【示例 8-4】调用 SQLiteDatabase 类的静态方法创建数据库。新建项目 SQLiteDatabase，在代码中调用 SQLiteDatabase.openOrCreateDatabase(String path,SQLiteDatabase.CursorFactory factory)静态方法创建数据库。

逻辑代码如下：

```
public class SQLiteDatabaseActivity extends Activity {
```

```
@Override
public void onCreate(Bundle savedInstanceState) {
    super.onCreate(savedInstanceState);
    setContentView(R.layout.activity_sqlite_database);
    //在 data/data/com.example.sqlitedatabase 目录下创建名为 users.db 的数据库
        SQLiteDatabase database = SQLiteDatabase.openOrCreateDatabase
                ("/data/data/com.example.sqlitedatabase/uesrs.db", null);
    }
}
```

运行程序，打开 DDMS 视图的 File Explorer 面板。在 data/data/com.example.sqlitedatabase 目录下，生成 users.db 数据库文件，表示数据库创建成功，如图 8.8 所示。

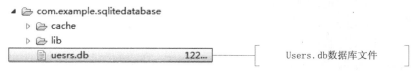

图 8.8 创建数据库 users.db

8.4.2 数据库编程操作

在得到数据库对象之后，便可以对数据库进行操作。SQLiteDatabase 类提供了一系列方法，实现数据库的增、删、改、查，如表 8-3 所示。

表 8-3 数据库基本操作

方 法 名 称	方 法 说 明
execSQL(String sql)	执行 SQL 语句
execSQL(String sql, Object[] bindArgs)	执行带占位符的 SQL 语句
insert(String table, String nullColumnHack, ContentValues values)	向表中插入一条记录
update(String table, ContentValues values, String whereClause, String[] whereArgs)	更新表中指定的某条记录
delete(String table, String whereClause, String[] whereArgs)	删除表中指定的某条记录
query(String table, String[] columns, String selection, String[] selectionArgs, String groupBy, String having, String orderBy)	查询表中记录
rawQuery(String sql, String[] selectionArgs)	查询带占位符的记录

表 8-3 中的查询方法，其返回值都是一个 Cursor 对象。Cursor 提供了以下方法移动查询记录的游标，如表 8-4 所示。

表 8-4 Cursor 移动方法

方 法 名 称	方 法 说 明
move(int offset)	从当前位置将游标向上或向下移动的行数。offest 为正数表示向下移，负数表示向上移
moveToFirst()	将游标移动到第一行，成功返回 true
moveToLast()	将游标移动到最后一行，成功返回 true
moveToNext()	将游标移动到下一行，成功返回 true
moveToPosition(int position)	将游标移动到指定行，成功返回 true
moveToPrevious()	将游标移动到上一行，成功返回 true

当游标移动到指定位置，就可以调用 Cursor 的 getXXX()方法，获取该行指定列的对应数据。

【示例 8-5】 数据库基本操作。在【示例 8-4】SQLiteDatabase 程序的界面上，添加 5 个按钮，分别用于创建表、插入记录、更新记录、查询记录、删除记录。

逻辑代码如下：

```java
public class SQLiteDatabaseActivity extends Activity {
    private Button create,insert,query,update,delete;
    @Override
    public void onCreate(Bundle savedInstanceState) {
        super.onCreate(savedInstanceState);
        setContentView(R.layout.activity_sqlite_database);
        //创建数据库
        final SQLiteDatabase database = SQLiteDatabase.openOrCreateDatabase
            ("/data/data/com.example.sqlitedatabase/uesrs.db", null);
        //创建 user_info 表
        create = (Button)findViewById(R.id.button1);
        create.setOnClickListener(new OnClickListener() {
            public void onClick(View v) {
                // TODO Auto-generated method stub
                //创建表语句
                String creatStr = "create table user_info(" +
                    "_id int," +
                    "name char(20)," +
                    "age int)";
                database.execSQL(creatStr);
            }
        });
        //插入两条记录
        insert = (Button)findViewById(R.id.button2);
        insert.setOnClickListener(new OnClickListener() {
            public void onClick(View v) {
                // TODO Auto-generated method stub
                //带占位符的插入语句
                String insertStr1 = "insert into user_info(_id,name,age) "+
                    " values(?,?,?)";
                //插入的数据
                Object[] valuesObjects = {1,"Seven",22};
                //执行带占位符的插入语句
                database.execSQL(insertStr1,valuesObjects);
                //不带占位符的插入语句
                String insertStr2 = "insert into user_info(_id,name,age) "+
                    "values(2,'Jim',24)";
                //执行不带占位符的插入语句
                database.execSQL(insertStr2);
            }
        });
        //更新记录
        update = (Button)findViewById(R.id.button3);
        update.setOnClickListener(new OnClickListener() {
            public void onClick(View v) {
                // TODO Auto-generated method stub
                //声明 ContentValues 对象
                ContentValues values = new ContentValues();
                //为 ContentValues 添加值
                values.put("name", "BOB");
                //更新_id 为 1 的记录 name 为 BOB
```

```java
            database.update("user_info", values, "_id=?", new
                String[]{"1"});
        }
    });
    //查询记录。在主界面添加 ListView,以列表形式显示记录
    query = (Button)findViewById(R.id.button4);
    query.setOnClickListener(new OnClickListener() {
        public void onClick(View v) {
            // TODO Auto-generated method stub
            //以年龄降序排列记录
            Cursor cursor = database.query("user_info", new String[]
            {"_id","name","age"}, null, null, null, null, "age desc");
            //声明适配器为 ListView 提供数据
            SimpleCursorAdapter sCursorAdapter = new SimpleCursorAdapter(
                //上下文环境
                SQLiteDatabaseActivity.this,
                // ListView 布局文件
                R.layout.users,
                //游标
                cursor,
                //表中的列名
                new String[]{"_id","name","age"},
                //ListView 显示内容
                new int[]{R.id.editText3,R.id.editText1,
                R.id.editText2},
                //设置 Cursor 监听
                CursorAdapter.FLAG_REGISTER_CONTENT_OBSERVER);
                //声明 ListView
            ListView listView = (ListView)findViewById
            (R.id.user_info);
            //绑定适配器
            listView.setAdapter(sCursorAdapter);
        }
    });
    //删除记录
    delete =(Button)findViewById(R.id.button5);
    delete.setOnClickListener(new OnClickListener() {
        public void onClick(View v) {
            // TODO Auto-generated method stub
            //删除_id 为 2 的记录
            database.delete("user_info", "_id=?", new String[]{"2"});
        }
    });
    }
}
```

查看表结构,需要下载安装辅助工具 SQLite Expert Personal 软件。导出 users.db 数据库文件,打开 SQLite Expert Personal,单击左上角的 File 按钮,在弹出的菜单中选择 Open Database 命令,打开 user_info 表。在 Data 面板中查看表结构,如图 8.9 所示。

8.4.3 SQLiteOpenHelper 类

Android 还提供了一个数据库辅助类——SQLiteOpenHelper。SQLiteOpenHelper 类根据开发应用程序的需要,封装了创建和更新数据库使用的逻辑。只要继承 SQLiteOpenHelper 类,并重写其中的 onCreate()和 onUpgrade()方法,就可以创建数据库。SQLiteOpenHelper

创建user_info表

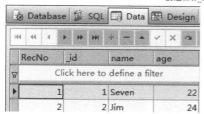
插入记录

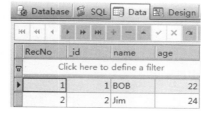
修改_id为1的记录name为BOB

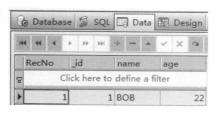
删除_id为2的记录

查询记录按年龄降序排列

图 8.9 数据库基本操作

使数据库的管理、维护和升级更加方便,它的构造方法语法如下:

```
SQLiteOpenHelper(Context context,      // 上下文环境
    String name,                        // 数据库文件名
    CursorFactory factory,              // 游标
    int version)                        // 数据库版本
```

SQLiteOpenHelper 类提供了以下列相关方法创建、打开或关闭数据库,如表 8-5 所示。

表 8-5　SQLiteOpenHelper 相关方法

方 法 名 称	方 法 说 明
close()	关闭数据库
getDatabaseName()	获取数据库名称
getReadableDatabase()	创建或打开一个只读数据库

方 法 名 称	方 法 说 明
getWritableDatabase()	创建或打开一个可读可写的数据库
onCreate(SQLiteDatabase db)	第一次创建数据库时调用
onOpen(SQLiteDatabase db)	打开数据库时调用
onUpgrade(SQLiteDatabase db, int oldVersion, int newVersion)	数据库版本升级时调用

【示例 8-6】 使用 SQLiteOpenHelper 类创建数据库。新建项目 SQLite。新建 DatabaseHelper 类继承于 SQLiteOpenHelper。在 SQLiteActivity 界面中添加两个按钮，一个用于创建只读数据库，另一个用于创建可读可写数据库。

DatabaseHelper：

```java
//新建DatabaseHelper类,继承于SQLiteOpenHelper
public class DatabaseHelper extends SQLiteOpenHelper {
    public DatabaseHelper(Context context, String name, CursorFactory factory,
            int version) {
        super(context, name, factory, version);
        // TODO Auto-generated constructor stub
    }
    @Override
    //覆盖onCreate()方法,第一次创建数据库时执行该方法
    public void onCreate(SQLiteDatabase arg0) {
        // TODO Auto-generated method stub
        System.out.println("数据库创建");
    }
    @Override
    //覆盖onUpgrade()方法,数据库需要升级时执行该方法
    public void onUpgrade(SQLiteDatabase db, int oldVersion, int newVersion){
        // TODO Auto-generated method stub
        System.out.println("数据库更新");
    }
}
```

SQLiteActivity：

```java
public class SQLiteActivity extends Activity {
    private Button b1,b2;
    @Override
    public void onCreate(Bundle savedInstanceState) {
        super.onCreate(savedInstanceState);
        setContentView(R.layout.activity_sqlite);
        //创建只读数据库
        b1 = (Button)findViewById(R.id.button1);
        b1.setOnClickListener(new OnClickListener() {

            public void onClick(View v) {
                // TODO Auto-generated method stub
                //创建DatabaseHelper对象
                DatabaseHelper dHelper = new DatabaseHelper(
                //当前上下文环境
                SQLiteActivity.this,
                //数据库名称
                "one.db",
                //使用默认的CursorFactory
```

```
                null,
                //数据库版本
                1);
                //创建只读数据库
                SQLiteDatabase db = dHelper.getReadableDatabase();
            }
        });
        //创建可读可写数据库
        b2 = (Button)findViewById(R.id.button2);
        b2.setOnClickListener(new OnClickListener() {
            public void onClick(View v) {
                // TODO Auto-generated method stub
                //创建 DatabaseHelper 对象
                DatabaseHelper dHelper = new DatabaseHelper(
                //当前上下文环境
                SQLiteActivity.this,
                //数据库名称
                "two.db",
                //使用默认的 CursorFactory
                null,
                //数据库版本
                1);
                //创建可读可写数据库
                SQLiteDatabase db = dHelper.getWritableDatabase();
            }
        });
    }
}
```

运行程序,先后单击 Readable 和 Writable 按钮创建数据库。打开 DDMS 视图中的 File Explorer 面板。在 data/data/com.example.sqlite（项目包名）/databases 路径下,生成两个数据库文件,如图 8.10 所示。

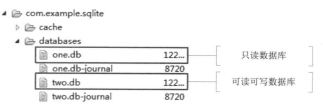

图 8.10　使用 SQLiteOpenHelper 创建数据库

8.5　数据共享 ContentPrivoder

ContentPrivoder 是所有应用程序进行数据存储的一个桥梁,它能使各个应用程序之间实现数据共享。ContentPrivoder 是 Android 提供的四大组件之一,在 Android 中可支持多个应用中存储和读取数据,这也是 Android 中跨应用共享数据的唯一方式。

8.5.1　ContentPrivoder 简介

一个程序可以通过实现一个 ContentProvider 的抽象接口,将自己的数据以 Uri 形式完全暴露出去。其他应用程序就可以使用 ContentResolver,根据 Uri 访问操作指定的数据。

1. Uri 简介

以下是一些示例 Uri。

- content://media/internal/images：将返回设备上存储的所有图片；
- content://contacts/people/：将返回设备上的所有联系人信息；
- content://contacts/people/45：返回单个结果（联系人信息中 ID 为 45 的联系人记录）。

Uri 工具类提供了静态方法 parse()，将字符串解析为 Uri，代码如下：

```
Uri uri = Uri.parse("content://Contacts/ people/45");
```

2. ContentProvider 相关方法

ContentProvider 也提供了一些方法，对数据进行增、删、改、查等操作，如表 8-6 所示。

表 8-6　ContentProvider 相关方法

方　法　名　称	方　法　说　明
delete(Uri uri, String selection, String[] selectionArgs)	删除一行或多行数据
insert(Uri uri, ContentValues values)	插入一行数据
query(Uri uri, String[] projection, String selection, String[] selectionArgs, String sortOrder)	查询数据
update(Uri uri, ContentValues values, String selection, String[] selectionArgs)	更新一个或多个数据

【示例 8-7】 启动模拟器，在模拟器的 Contacts 中手动添加几条联系人信息。新建项目 Contacts，使用 ContentProvider 访问设备上存储的联系人信息。

逻辑代码如下：

```java
public class ContactsActivity extends Activity {
    private ContentResolver cResolver;

    @Override
public void onCreate(Bundle savedInstanceState) {
        //声明 ContentResolver 对象
        super.onCreate(savedInstanceState);
        setContentView(R.layout.activity_contacts);
        //获取 ContentResolver 实例对象
        cResolver = getContentResolver();
        //获取联系人 Uri
        Uri uri = ContactsContract.Contacts.CONTENT_URI;
        //联系人姓名
        String name = "wyl";
        //联系人电话号码
        String phone1 = "1111111";
        //空插入,获取系统返回的 raw_contact_id
        Uri rawUri = ContactsContract.RawContacts.CONTENT_URI;
        ContentValues values = new ContentValues();
        Uri insertuUri = cResolver.insert(rawUri, values);
        long raw_contact_id = ContentUris.parseId(insertuUri);

        //插入姓名
        Uri dataUri = ContactsContract.Data.CONTENT_URI;
        values.clear();
```

```java
values.put(ContactsContract.Data.RAW_CONTACT_ID,raw_contact_id);
values.put(ContactsContract.Data.MIMETYPE, StructuredName.
CONTENT_ITEM_TYPE);
values.put(StructuredName.DISPLAY_NAME, name);
cResolver.insert(dataUri, values);

//插入电话号码
values.clear();
values.put(ContactsContract.Data.RAW_CONTACT_ID, raw_contact_id);
values.put(ContactsContract.Data.MIMETYPE, Phone.CONTENT_
ITEM_TYPE);
values.put(Phone.TYPE, Phone.TYPE_MOBILE);
values.put(Phone.NUMBER, phone1);
cResolver.insert(ContactsContract.Data.CONTENT_URI, values);

//修改联系人信息
values.clear();
values.put(ContactsContract.Data.RAW_CONTACT_ID, raw_contact_id);
//修改号码"1111111"为"3333333"
values.put(Phone.NUMBER,"3333333");
String where = Phone.NUMBER + "=1111111";
cResolver.update(dataUri, values, where, null);

//删除联系人信息
String where1 = Phone.NUMBER + "=3333333";
//删除号码为"3333333"的联系人
cResolver.delete(dataUri, where1, null);

//查询联系人信息
Cursor cursor = cResolver.query(uri, null, null, null, null);

//游标移动到下一行
while (cursor.moveToNext()) {
  //联系人姓名
  String display_name = cursor.getString(cursor.getColumnIndex(
      ContactsContract.Contacts.DISPLAY_NAME));
//联系人ID
  String contact_id = cursor.getString(cursor.getColumnIndex(
      ContactsContract.Contacts._ID));
//联系人电话Uri
  Uri pUri = ContactsContract.CommonDataKinds.Phone.CONTENT_URI;
// Where 子句
  String selection = ContactsContract.CommonDataKinds.Phone.
CONTACT_ID + "=" + contact_id;
//查询电话号码
  Cursor phoneCursor = cResolver.query(pUri, null, selection, null, null);
  while (phoneCursor.moveToNext()) {
      //获取电话号码
      String phone = phoneCursor.getString(phoneCursor.
getColumnIndex(
          ContactsContract.CommonDataKinds.Phone.DATA));
      //输出联系人信息
      System.out.println("联系人姓名: " + display_name + "\n号
        码: " + phone);
  }
 }
}
}
```

AndroidManifest.xml：

```xml
<manifest xmlns:android="http://schemas.android.com/apk/res/android"
    package="com.example.contacts"
    android:versionCode="1"
    android:versionName="1.0" >
    <uses-sdk
        android:minSdkVersion="8"
        android:targetSdkVersion="15" />

    <!--添加用户权限允许程序读写联系人数据-- >
    <uses-permission android:name="android.permission.READ_CONTACTS"/>
    <uses-permission android:name="android.permission.WRITE_CONTACTS"/>

    <application
        android:icon="@drawable/ic_launcher"
        android:label="@string/app_name"
        android:theme="@style/AppTheme" >
        <activity
            android:name=".ContactsActivity"
            android:label="@string/title_activity_contacts" >
            <intent-filter>
                <action android:name="android.intent.action.MAIN" />
                <category android:name="android.intent.category.LAUNCHER" />
            </intent-filter>
        </activity>
    </application>

</manifest>
```

程序运行结果如图 8.11 所示。

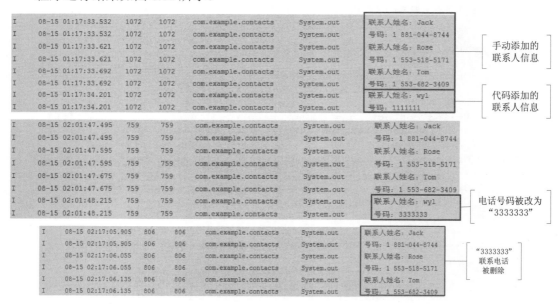

图 8.11 程序运行结果

8.5.2 ContentProvider 的应用

前面介绍的是如何使用 ContentResolver 来操作系统 ContentProvider 提供的数据。本节我们将详细介绍如何实现自定义的 ContentProvider。

1. 创建 ContentProvider 的步骤

（1）创建一个类继承于 ContentProvider 父类，该类实现增、删、改、查等方法；

（2）在自定义的 ContentProvider 子类中，定义一个 public static final 的名为 CONTENT_URI 的 Uri 类变量。为其指定一个唯一的字符串值，最好的方案是设定为类的全名称。如 public static final Uri CONTENT_URI = Uri.parse("content://com.example.MyContentProvider");；

（3）在 AndroidMenifest.xml 中使用<provider...></provider>标签来设置 ContentProvider，并设置 "android:authorities" 属性。语法如下：

```
<provider
        android:authorities=" "      <!--CONTENT_URI 的 authority 值-->
        android:name=" ">            <!--自定义 ContentProvider 名称-->
</provider>
```

2. 工具类

开发自定义的 ContentProvider 类时所实现的增、删、改、查方法，都需要一个 Uri 参数。该参数决定对哪个 Uri 进行操作。

为了确定该 ContentProvider 实际能匹配的 Uri，以及每个方法中 Uri 参数所操作的数据，Android 系统提供了 UriMatcher 类。UriMatcher 类的相关方法如表 8-7 所示。

表 8-7　UriMatcher 类的相关方法

方 法 名 称	方 法 说 明
addURI(String authority, String path, int code)	增加一个 Uri 去匹配。authority 和 path 组成一个 Uri，code 是匹配成功后返回的代码，必须是正数
match(Uri uri)	匹配 Uri，匹配不成功返回-1

除了 UriMatcher 之外，Android 系统还提供了 ContentUris 工具类，用于操作 Uri 字符串。ContentUris 的相关方法如表 8-8 所示。

表 8-8　ContentUris 的相关方法

方 法 名 称	方 法 说 明
appendId(Uri.Builder builder, long id)	将给定的 ID 加到路径末端
parseId(Uri contentUri)	解析 Uri 中包含的 ID 值
withAppendedId(Uri contentUri, long id)	为路径加上 ID 部分

3. 实现自定义 ContentProvider

开发一个 BookContentProvider，使用自定义的 ContentProvider 访问 Book 数据。

【示例 8-8】 自定义 ContentProvider 的开发。新建项目 ContentProvider。在程序中创建 BOOK 数据库和 BOOK 表。然后创建 BookContentProvider 继承于 ContentProvider，访问 Book 数据。

AndroidMenifest.xml：

```
<manifest xmlns:android="http://schemas.android.com/apk/res/android"
    package="com.example.contentprivoder"
    android:versionCode="1"
```

```xml
    android:versionName="1.0" >
<uses-sdk
    android:minSdkVersion="8"
    android:targetSdkVersion="15" />

<application
    android:icon="@drawable/ic_launcher"
    android:label="@string/app_name"
    android:theme="@style/AppTheme" >
    <activity
        android:name=".ContentPrivoderActivity"
        android:label="@string/title_activity_content_privoder" >
        <intent-filter>
            <action android:name="android.intent.action.MAIN" />
            <category android:name="android.intent.category.LAUNCHER" />
        </intent-filter>
    </activity>

    <provider
    <!-- AUTHORITY 值-- >
        android:authorities="content.provider.bookContent"
    <!--自定义 ContentProvider 名称-->
        android:name="BookContentPrivoder">

    </provider>
</application>
</manifest>
```

BookContentProvider：

```java
public class BookContentPrivoder extends ContentProvider {
    // AUTHORITY 值
    private final static String AUTHORITY = "content.provider.bookContent";
    //定义该 Content 提供的两个 Uri
    public final static String CONTENT_URI = "content:// "+AUTHORITY+
    "/books";
    public final static String CONTENT_URI_BOOK = "content:// "+AUTHORITY+
    "/book";
    //数据表的字段名称,_id,title,isbn
    public final static String _ID = "_id";
    public final static String TITLE = "title";
    public final static String ISBN = "isbn";
    //Uri 的注册 code
    public final static int BOOKS = 1;
    public final static int BOOK_ID = 2;
    private static UriMatcher uriMatcher = new UriMatcher(UriMatcher.
    NO_MATCH);
    //匹配 Uri
        static{
            uriMatcher.addURI(AUTHORITY, "books", BOOKS);
            uriMatcher.addURI(AUTHORITY, "book/#", BOOK_ID);
        }
    //数据库名称
    private static String DATABASE_NAME = "books.db";
    //数据表名称
    private static String TABLE_NAME = "booksTable";
    //创建表语句
```

```java
    private static String DATABASE_CREATE = "create table "
        + TABLE_NAME + " (_id integer primary key autoincrement, "
        + "title text not null, isbn text not null);";
    private SQLiteDatabase bookDb;
    private class dbHelper extends SQLiteOpenHelper{
        public dbHelper(Context context, String name,
                CursorFactory factory, int version) {
            super(context, name, factory, version);
            // TODO Auto-generated constructor stub
        }
        @Override
        public void onCreate(SQLiteDatabase db) {
            // TODO Auto-generated method stub
            //创建表
            db.execSQL(DATABASE_CREATE);
        }
        @Override
        public void onUpgrade(SQLiteDatabase db, int oldVersion, int newVersion) {
            // TODO Auto-generated method stub
        }
    }
    @Override
    public int delete(Uri uri, String selection, String[] selectionArgs) {
        // TODO Auto-generated method stub
        return 0;
    }
    @Override
    public String getType(Uri uri) {
        // TODO Auto-generated method stub
        return null;
    }
    @Override
    public Uri insert(Uri uri, ContentValues values) {
        // TODO Auto-generated method stub
        //插入book信息
        long rowid = bookDb.insert(TABLE_NAME, "title", values);
        Uri uriInsert = ContentUris.withAppendedId(uri, rowid);
        getContext().getContentResolver().notifyChange(uri, null);
        return uriInsert;
    }
    @Override
    public boolean onCreate() {
        // TODO Auto-generated method stub
        //创建数据库
        dbHelper dbHelperObj = new dbHelper(getContext(), DATABASE_NAME,
        null, 1);
        bookDb = dbHelperObj.getWritableDatabase();
        return (bookDb==null)?false:true;
    }
    @Override
    //查询book信息
    public Cursor query(Uri uri, String[] projection, String selection,
            String[] selectionArgs, String sortOrder) {
        // TODO Auto-generated method stub
        Cursor cursor = null;
        cursor = bookDb.query(TABLE_NAME, projection, selection,
        selectionArgs, null, null, sortOrder);
        return cursor;
    }
```

```java
    @Override
    public int update(Uri uri, ContentValues values, String selection,
            String[] selectionArgs) {
        // TODO Auto-generated method stub
        return 0;
    }
}
```

ContentProviderActivity：

```java
public class ContentPrivoderActivity extends Activity {
    //声明 ContentResolver 对象
    private ContentResolver contentResolver;
    @Override
    public void onCreate(Bundle savedInstanceState) {
        super.onCreate(savedInstanceState);
        setContentView(R.layout.activity_content_privoder);
        //获取 ContentResolver 实例对象
        contentResolver = getContentResolver();
        //获取数据库 Uri
        Uri uri = Uri.parse(BookContentPrivoder.CONTENT_URI);
        //插入 book 信息
        ContentValues values = new ContentValues();
        values.put(BookContentPrivoder.TITLE, "Android4.1");
        values.put(BookContentPrivoder.ISBN, "0000-5677-7651");
        Uri insertUri = contentResolver.insert(uri, values);
        //查询 book 信息
        long rowid = ContentUris.parseId(insertUri);
        Uri bookUri = ContentUris.withAppendedId(Uri.parse(
            BookContentPrivoder.CONTENT_URI_BOOK), rowid);
        System.out.println("bookUri="+bookUri);
        String[] projection = {BookContentPrivoder.TITLE,
        BookContentPrivoder.ISBN};
        Cursor cursor = contentResolver.query(bookUri, projection, null,
        null, null);
        while (cursor.moveToNext()) {
            String title = cursor.getString(0);
            String isbn = cursor.getString(1);
            System.out.println("title="+title+";ISBN="+isbn);
        }
    }
}
```

运行程序，查看 LogCat 面板的输出信息。如图 8.12 所示。

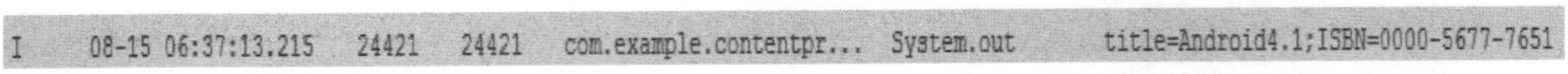

`I  08-15 06:37:13.215   24421    24421   com.example.contentpr... System.out       title=Android4.1;ISBN=0000-5677-7651`

图 8.12 LogCat 面板输出的信息

在辅助工具 SQLite Expert Personal 软件中查看 Book 信息，如图 8.13 所示。

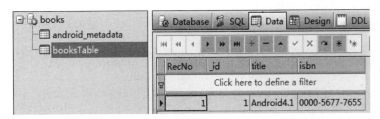

图 8.13 自定义 ContentProvider

8.6 小　　结

本章主要讲解了 Android 平台下的数据存储，详细介绍了 4 种存储方式的使用方法以及适用情况。其中，SharedPreferences 和文件存取是较为简单的存储方式，容易掌握；SQLite 数据库存储在 Android 程序中较为常用；ContentPrivoder 是本章的难点，需要读者细心学习，认真掌握。

8.7 习　　题

1. 新建项目 SharedPreferences，使用 SharedPreferences 实现数据存储。在 File Explorer 面板中生成 file.xml 文件，导出查看，如图 8.14 所示。

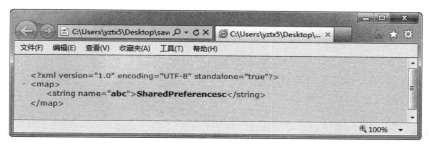

图 8.14　SharedPreferences

【分析】本题目考查了读者对 SharedPreferences 存储的掌握。可以参考 8.1 节的内容。
【核心代码】本题的核心代码如下所示。

```
SharedPreferences spf = getSharedPreferences("save",MODE_WORLD_
READABLE);
 SharedPreferences.Editor edit = spf.edit();
 edit.putString("abc", "Sharedpreferences");
 edit.commit();
```

2. 新建项目 SQLiteDatabase。在代码中调用 SQLiteDatabase.openOrCreateDatabase(String path,SQLiteDatabase.CursorFactory factory) 静态方法创建数据库，然后对数据库进行增、删、改、查等操作。数据库创建成功后，在 data/data/com.example.sqlitedatabase 目录下，生成 students.db 数据库文件，如图 8.15 所示。

图 8.15　创建数据库

【分析】本题目考查了读者对使用静态方法创建数据库，以及数据库的基本操作的掌握。可以参考 8.3 节的内容。
【核心代码】本题的核心代码如下所示。

```
SQLiteDatabase database = SQLiteDatabase.openOrCreateDatabase
        ("/data/data/com.example.sqlitedatabase/students.db", null);

    insert = (Button)findViewById(R.id.button2);
    insert.setOnClickListener(new OnClickListener() {
```

```java
        public void onClick(View v) {
            // TODO Auto-generated method stub
            String insertStr1 = "insert into user_info(_id,name,age)"+
                    "values(?,?,?)";
            Object[] valuesObjects = {1,"Seven",22};
            database.execSQL(insertStr1,valuesObjects);
            
            String insertStr2 = "insert into user_info(_id,name,age)"+
                    "values(2,'Jim',24)";
            database.execSQL(insertStr2);
        }
    });
    
    update = (Button)findViewById(R.id.button3);
    update.setOnClickListener(new OnClickListener() {
        
        public void onClick(View v) {
            // TODO Auto-generated method stub
            ContentValues values = new ContentValues();
            values.put("name", "BOB");
            database.update("user_info", values, "_id=?", new
                    String[]{"1"});
        }
    });
    
    query = (Button)findViewById(R.id.button4);
    query.setOnClickListener(new OnClickListener() {
        
        public void onClick(View v) {
            // TODO Auto-generated method stub
            Cursor cursor = database.query("user_info", new String[]
            {"_id","name","age"},
                    null, null, null, null, "age desc");
            SimpleCursorAdapter sCursorAdapter = new SimpleCursorAdapter(
                    SQLiteDatabaseActivity.this,
                    R.layout.users,
                    cursor,
                    new String[]{"_id","name","age"},
                    new int[]{R.id.editText3,R.id.editText1,
                    R.id.editText2},
                    CursorAdapter.FLAG_REGISTER_CONTENT_OBSERVER);
            ListView listView = (ListView)findViewById(R.id.user_info);
            listView.setAdapter(sCursorAdapter);
        }
    });
    
    delete =(Button)findViewById(R.id.button5);
    delete.setOnClickListener(new OnClickListener() {
        
        public void onClick(View v) {
            // TODO Auto-generated method stub
            database.delete("user_info", "_id=?", new String[]{"2"});
        }
    });
```

第 2 篇　Android 典型应用与实战

- 第 9 章　Android 网络应用
- 第 10 章　Android 中图形图像的处理
- 第 11 章　Android 多媒体应用
- 第 12 章　Android 感应检测——Sensor
- 第 13 章　手势识别和无线网络
- 第 14 章　Google 地图服务
- 第 15 章　Android 通信服务
- 第 16 章　Android 特色应用开发
- 第 17 章　Android 应用开发——网上购书

第 9 章 Android 网络应用

如今，随着现代网络的发展，互联网在手机中的应用发挥了巨大的作用，我们可以无线上网、可以进行视频通话、可以浏览网页等。Android 是由互联网巨头 Google 带头开发，因此对网络功能的支持是必不可少的。在 Android 系统中，提供了以下几种方式可以实现网络通信：Socket 通信、HTTP 通信、URL 通信、WebView 网络开发。本章将详细介绍这几种通信方式。

9.1 Socket 网络通信

Socket，通常也称作"套接字"，用于描述 IP 地址和端口。应用程序通常通过"套接字"向网络发出请求或者应答网络请求。Socket 是 Java 中较为常用的网络通信方式，而 Android 是采用 Java 语言进行开发。因此 Android 中 Socket 通信，采用的就是 Java 的 Socket 通信方式。

9.1.1 Socket 工作机制

Socket 工作机制中包括服务端和客户端两部分。在服务端有多个端口，每个端口由端口号标识。当客户端与服务端要建立连接时，首先服务端打开端口监听来自客户端的请求，接着客户端通过 IP 地址和端口号向服务端发送连接请求，然后服务端接收请求，则连接成功，便可以开始进行通信。工作模式如图 9.1 所示。

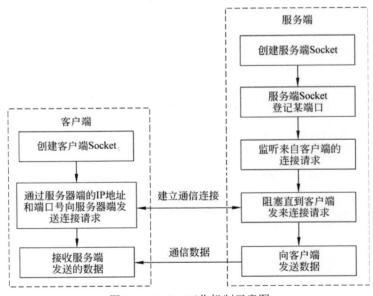

图 9.1 Socket 工作机制示意图

Socket 机制用到的类有 java.net.ServerSocket、java.net.Socket 等。服务器端以监听端口号为参数实例化 ServerSocket 类，以 accept()方法接收客户的连接。

```
ServerSocket ss = new ServerSocket(Int port);
Socket socket = ss.accept();
```

其中，ss 是声明一个 ServerSocket 对象；ServerSocket()方法创建一个新的 ServerSocket 对象并绑定到给定端口；accept()方法用来接受客户连接。

客户端则直接以服务器的地址和监听端口为参数实例化 Socket 类，连接服务器。

```
Socket socket = new Socket(String dstName,int dstPort);
```

当两者建立连接后，就可以进行网络通信。服务器端和客户端之间是通过流的形式进行交互。服务端调用 getOutputStream()方法得到输出流，并向其中写入数据信息传递给客户端。

```
DataOutputStream dout = new DataOutputStream(socket.getOutputStream());
```

客户端调用 getInputStream()方法得到输入流，接收服务端发送的数据信息。

```
DataInputStream din = new DataInputStream(socket.getInputStream());
```

9.1.2 Socket 服务端

Socket 服务端用于向客户端发送数据信息，它运行在 Java SE 平台上。

【示例 9-1】 Socket 服务端程序的开发。新建一个 Java Project，命名为 Server，步骤如图 9.2 所示。

然后在 Server 的 src 目录下新建一个包，命名为 com.example.server，步骤如图 9.3 所示。

在 com.example.server 包下新建一个 Server.java，添加 main()方法。在 main()方法中添加代码，向数据流中写入数据，并发送到客户端。

Server.java：

```
public class Server {
    public static void main(String[] args) {
        try {
            //声明一个服务端口8888
            ServerSocket ss = new ServerSocket(8888);
            //打印信息提示等待连接
            System.out.println("Listening... ");
            while (true) {
            //打开连接等待请求传入
                Socket socket = ss.accept();
                //打印信息提示已与客户端连接成功
                System.out.println("Cilent Connected... ");
                //获得输出流
                DataOutputStream dout = new DataOutputStream(socket.getOutputStream());
                //输出流数据
                String str = "Socket通信";
                //将数据写入到输出流
```

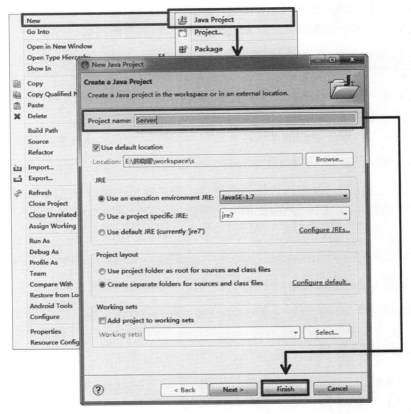

图 9.2 新建 Java Project

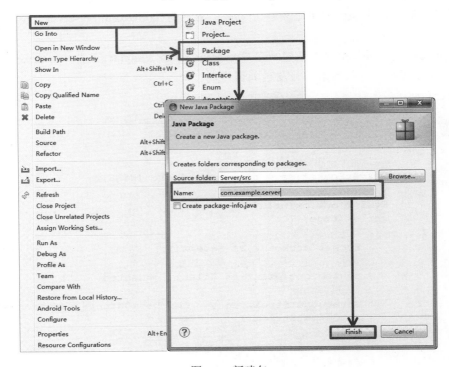

图 9.3 新建包

```
            dout.writeUTF(str);
            //关闭输出流
            dout.close();
            //关闭socket
            socket.close();
        }
    } catch (IOException e) {
        // TODO Auto-generated catch block
        e.printStackTrace();
    }
  }
}
```

9.1.3 Socket 客户端

Socket 客户端用于接收服务端发送的数据,运行在 Android 平台上。

【示例 9-2】 新建项目 Socket。在 SocketActivity 中创建 connectToServer()方法。然后添加一个按钮,单击按钮开始通信服务,读取服务端发送的信息,并用 Toast 显示。

逻辑代码如下:

```
@TargetApi(11)
public class SocketActivity extends Activity {
    //连接按钮
    private Button  button ;
    @Override
    public void onCreate(Bundle savedInstanceState) {
        super.onCreate(savedInstanceState);
        setContentView(R.layout.activity_socket);
        //添加以下两段代码,防止程序抛出 NetworkOnMainThreadException 异常
        StrictMode.setThreadPolicy(new StrictMode.ThreadPolicy.Builder()
        //设置线程策略
        .detectDiskWrites()
        .detectDiskReads()
        .detectNetwork()
        .penaltyLog()
        .build());
        StrictMode.setVmPolicy(new StrictMode.VmPolicy.Builder()
        //设置虚拟内存策略
        .detectLeakedSqlLiteObjects()
        .detectLeakedClosableObjects()
        .penaltyLog()
        .penaltyDeath()
        .build());
        button = (Button)findViewById(R.id.button1);
        //绑定监听,单击按钮发送连接请求
        button.setOnClickListener(new OnClickListener() {
            public void onClick(View v) {
                // TODO Auto-generated method stub
        //调用 connectToServer()方法
                connetTOServer();
            }
        });
}
        //创建 connectToServer()方法
        private void connetTOServer() {
            // TODO Auto-generated method stub
            try {
                //创建 socket 连接
```

```
            Socket socket = new Socket("192.168.0.103", 8888);
            //获得输入流
            DataInputStream din = new DataInputStream(socket.
            getInputStream());
            //读取服务端发送的信息
            String msg = din.readUTF();
            // Toast 显示信息
            Toast.makeText(SocketActivity.this, msg, Toast.LENGTH_LONG).
            show();
        } catch (UnknownHostException e) {
            // TODO Auto-generated catch block
            e.printStackTrace();
        } catch (IOException e) {
            // TODO Auto-generated catch block
            e.printStackTrace();
        }
    }
}
```

注意：在创建 Socket 连接时传入的 IP 地址为本示例测试时的主机地址，读者在自行测试时请修改为对应主机的 IP 地址。

AndroidMenifest.xml：

```
<manifest xmlns:android="http://schemas.android.com/apk/res/android"
    package="com.example.socket"
    android:versionCode="1"
    android:versionName="1.0" >
    <uses-sdk
        android:minSdkVersion="8"
        android:targetSdkVersion="15" />

    <uses-permission android:name="android.permission.INTERNET" /><!--允许
    应用程序访问网络-->

    <application
        android:icon="@drawable/ic_launcher"
        android:label="@string/app_name"
        android:theme="@style/AppTheme" >
        <activity
            android:name=".SocketActivity"
            android:label="@string/title_activity_socket" >
            <intent-filter>
                <action android:name="android.intent.action.MAIN" />
                <category android:name="android.intent.category.LAUNCHER" />
            </intent-filter>
        </activity>
    </application>
</manifest>
```

9.1.4 Socket 通信

完成了服务端与客户端的开发，下面运行程序进行 Socket 通信。

1. 运行 Socket 服务端

右击 Server.java，依次选择 Run As|Java Application 命令。服务端成功启动后，控制台输出提示信息"Listening…"，如图 9.4 所示。

第 9 章　Android 网络应用

图 9.4　服务端成功启动

2. 运行 Android 客户端

运行 SocketActivity，启动客户端。单击"连接"按钮，客户端发送连接请求。服务端接收到请求后，在控制台输出提示信息"Client Connected…"，连接成功。然后客户端就可以读取到服务端发送的信息，并将信息以 Toast 方式显示在界面上，如图 9.5 所示。

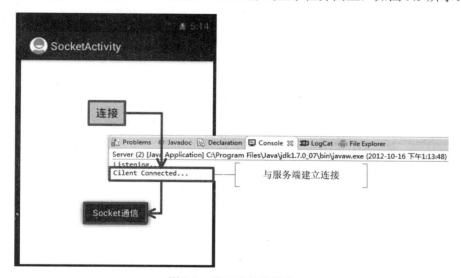

图 9.5　获取服务端信息

9.2　HTTP 网络通信

HTTP（Hyper Text Transport Protocol）超文本传送协议是一种通信协议。它用来传输超文本的数据，目前我们访问的大多数网页使用的就是 HTTP 网络通信协议。Android 提供了 HttpURLConnection 和 HttpClient 两个接口开发访问网站的程序。

9.2.1　HTTP 通信方式

HTTP 详细规定了浏览器和万维网（World Wide Web）服务器之间互相通信的规则。客户机和服务器必须都支持 HTTP，才能在万维网上发送和接收 HTML 文档并进行交互。

HTTP 包含了 GET 和 POST 两种请求网络资源方式。GET 可以获得静态页面，也可以把参数放在 URL 字符串后面，传递给服务器。而 POST 方法的参数是放在 HTTP 请求中。因此，在编程之前，应当首先明确使用的请求方法，然后再根据所使用的请求数据方法，选择相应的编程方式。

Android 提供了 HttpURLConnection 和 HttpClient 接口来开发 HTTP 程序。

9.2.2 HttpURLConnection 开发

HttpURLConnection 是 Java 的标准类，继承自 HttpConnection。它是一个抽象类，不能实例化对象，主要是通过 URL 的 openConnection 方法获得。语法如下：

```
URL url = new URL(" ");
HttpURLConnection conn = (HttpURLConnection)url.openConnection();
```

由于 openConnection()方法返回值类型是 URLConnection 类，所以需要强制转换类型为 HttpURLConnection 类。

openConnection()方法只创建 HttpURLConnection 实例，并不是真正的连接操作。而且每次调用 openConnection()方法，都将创建一个新的实例。因此，在连接之前可以对其一些属性进行设置。

- conn.setDoInput(true)：设置输入流；
- conn.setDoOutput(true)：设置输出流；
- conn.setConnectTimeout(10000)：设置超时时间；
- conn.setRequestMethod("GET")：设置请求方式，HttpURLConnection 默认使用 GET 方式；
- conn.setUseCaches(false)：POST 请求不能使用缓存。

【示例 9-3】新建项目 HttpURLConnection。使用 HttpURLConnection 的默认请求方式开发 HTTP 程序，请求网络数据。

逻辑代码：

```
@TargetApi(11)
public class HttpURLConnectionActivity extends Activity {
    @Override
    public void onCreate(Bundle savedInstanceState) {
        super.onCreate(savedInstanceState);
        setContentView(R.layout.activity_http_urlconnection);
      //添加以下两段代码,防止程序抛出 NetworkOnMainThreadException 异常
        StrictMode.setThreadPolicy(new StrictMode.ThreadPolicy.Builder()
        .detectDiskWrites()
        .detectDiskReads()
        .detectNetwork()
        .penaltyLog()
        .build());

        StrictMode.setVmPolicy(new StrictMode.VmPolicy.Builder()
        .detectLeakedSqlLiteObjects()
        .detectLeakedClosableObjects()
        .penaltyLog()
        .penaltyDeath()
        .build());
        try {
          //百度 Uri
            URL url = new URL("http://www.baidu.com/");
            //创建 HttpURLConnection 实例对象
            HttpURLConnection conn = (HttpURLConnection) url.openConnection();
            //获取输入流
            InputStreamReader in = new InputStreamReader(conn.getInputStream());
```

```
            //创建 BufferedReader
            BufferedReader buffer = new BufferedReader(in);
            String inputLine = null;
            String resultData = null;
            //使用循环读取数据
            while (((inputLine = buffer.readLine()) != null)) {
              resultData += inputLine + "\n";
            }
                //关闭输入流
                in.close();
                //关闭 HTTP 连接
                conn.disconnect();
                //打印网络数据
                System.out.println(resultData);
        } catch (IOException e) {
            // TODO Auto-generated catch block
            e.printStackTrace();
        }
    }
}
```

在 AndroidMenifest.xml 中添加用户权限，允许应用程序访问网络。

```
<uses-permission android:name="android.permission.INTERNET" />
```

运行程序，在 LogCat 面板中查看获取百度首页源文件，如图 9.6 所示为部分截图。

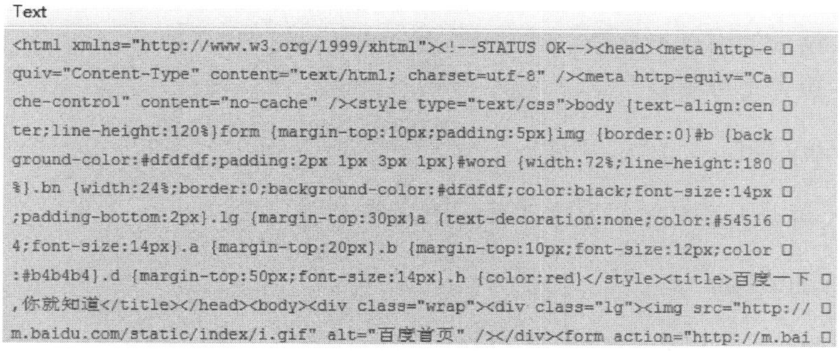

```
Text
<html xmlns="http://www.w3.org/1999/xhtml"><!--STATUS OK--><head><meta http-e
quiv="Content-Type" content="text/html; charset=utf-8" /><meta http-equiv="Ca
che-control" content="no-cache" /><style type="text/css">body {text-align:cen
ter;line-height:120%}form {margin-top:10px;padding:5px}img {border:0}#b {back
ground-color:#dfdfdf;padding:2px 1px 3px 1px}#word {width:72%;line-height:180
%}.bn {width:24%;border:0;background-color:#dfdfdf;color:black;font-size:14px
;padding-bottom:2px}.lg {margin-top:30px}a {text-decoration:none;color:#54516
4;font-size:14px}.a {margin-top:20px}.b {margin-top:10px;font-size:12px;color
:#b4b4b4}.d {margin-top:50px;font-size:14px}.h {color:red}</style><title>百度一下
,你就知道</title></head><body><div class="wrap"><div class="lg"><img src="http://
m.baidu.com/static/index/i.gif" alt="百度首页" /></div><form action="http://m.bai
```

图 9.6　获取百度首页源文件

9.2.3 HttpClient 接口开发

使用 Apache 提供的 HttpClient 接口同样可以进行 HTTP 操作。HttpClient 对 java.net 的类做了封装和抽象，更适合在 Android 上开发应用。在使用 HttpClient 接口开发 HTTP 时，会用到以下接口和类，下面将一一介绍。

1. HttpClient 接口

Http 客户端接口，DefaultHttpClient 是常用于实现 HttpClient 接口的子类。HttpClient 提供的抽象方法如表 9-1 所示。

2. HttpResponse 接口

Http 响应接口，HttpResponse 提供了一系列 get 方法，如表 9-2 所示。

表 9-1　HttpClient 接口的常用抽象方法

方 法 名 称	方 法 说 明
public abstract HttpResponse execute (HttpUriRequest request)	通过 HttpUriRequest 对象执行返回一个 HttpResponse 对象
public abstract HttpResponse execute (HttpUriRequest request, HttpContext context)	通过 HttpUriRequest 对象和 HttpContext 对象执行返回一个 HttpResponse 对象

表 9-2　HttpResponse 常用方法

方 法 名 称	方 法 说 明
public abstract HttpEntity getEntity()	得到一个 HttpEntity 对象
public abstract StatusLine getStatusLine()	得到一个 StatusLine 接口的实例对象
public abstract Locale getLocale()	得到 Locale 对象

3. StatusLine 接口

StatusLine 也就是 HTTP 协议中的状态行。HTPP 状态行由 3 部分组成：HTTP 协议版本、服务器发回的响应状态码、状态码的文本描述。

StatusLine 的子类 BasicStatusLine 类，提供了 public abstract int getStatusCode ()方法获得响应状态码。常见的响应状态码介绍如下。

- 200：服务器成功返回网页。
- 404：请求的网页不存在。
- 503：服务不可用。

4. HttpEntity 接口

HttpEntity 就是 HTTP 消息发送或接收的实体。

5. NameValuePair

NameValuePair 接口是一个简单的封闭的键值对。只提供了一个 getName()和一个 getValue()方法。主要用到的实现类是 BasicNameVaulePair 类。

6. HttpGet 类

HttpGet 实现了 HttpRequest、HttpUriRequest 接口，其构造方法如表 9-3 所示。

表 9-3　HttpGet 构造方法

方 法 名 称	方 法 说 明
public HttpGet ()	无参数构造方法用以实例化对象
public HttpGet (URI uri)	通过 URI 对象构造 HttpGet 对象
public HttpGet (String uri)	通过指定的 uri 字符串地址构造实例化 HttpGet 对象

7. HttpPost 类

HttpPost 也实现了 HttpRequest、HttpUriRequest 接口，其构造方法如表 9-4 所示。

表 9-4　HttpGet 构造方法

方 法 名 称	方 法 说 明
public HttpPost()	无参数构造方法用以实例化对象
public HttpPost (URI uri)	通过 URI 对象构造 HttpPost 对象
public HttpPost(String uri)	通过指定的 uri 字符串地址构造实例化 HttpPost 对象

掌握了以上这些 API 应用之后，我们就可以开发 Http 程序。使用 HttpClient 接口开发 HTTP 程序分为以下几个步骤：

（1）创建 HttpGet 或者 HttpPost 对象，将要请求的 URL 对象构造方法传入 HttpGet、HttpPost 对象。

（2）将第（1）步创建好的 HttpGet 对象或者 HttpPost 对象，传入 HttpClient 接口的实现类——DefaultHttpClent.excute(HttpUriRequest request)方法中，得到 HttpResponse 对象。

（3）通过 HttpResponse 提取到网络返回的一些信息，再做提取显示。

如图 9.7 所示为使用 GET 方式和 POST 方式获取网络资源信息的工作流程图。

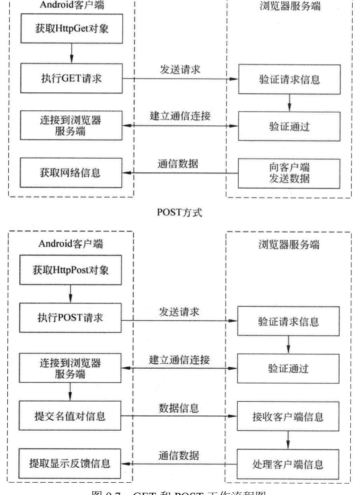

图 9.7　GET 和 POST 工作流程图

【示例 9-4】 使用 HttpClient 接口开发 HTTP 程序，使用 GET 方法获取网络资源信息。新建项目 HttpClient。在界面中添加 GET 按钮，用于发送 GET 请求。再添加一个空白文本框，用于显示返回的结果数据。并且在显示 GET 请求结果的文本框外添加 ScollView 控件，以便显示全部数据。

逻辑代码如下：

```java
@TargetApi(11)
public class HttpClientActivity extends Activity {
    //GET 按钮
    private Button get
    //显示静态网页数据
    private TextView geTextView
    @Override
    public void onCreate(Bundle savedInstanceState) {
        super.onCreate(savedInstanceState);
        setContentView(R.layout.activity_http_client);

     //添加以下两段代码,防止程序抛出 NetworkOnMainThreadException 异常
     StrictMode.setThreadPolicy(new StrictMode.ThreadPolicy.Builder()
     .detectDiskWrites()
     .detectDiskReads()
     .detectNetwork()
     .penaltyLog()
     .build());

     StrictMode.setVmPolicy(new StrictMode.VmPolicy.Builder()
     .detectLeakedSqlLiteObjects()
     .detectLeakedClosableObjects()
     .penaltyLog()
     .penaltyDeath()
     .build());
     get = (Button)findViewById(R.id.button1);
     get.setOnClickListener(new OnClickListener() {

        public void onClick(View v) {
            // TODO Auto-generated method stub
            try {
                //以人人网 IP 地址,实例化 HttpGet 对象
                HttpGet httpGet = new HttpGet("http://www.renren.com");
                //实现 HttpClient 接口
                HttpClient httpclient = new DefaultHttpClient();
                //声明 HttpResponse 对象
                HttpResponse hResponse;
                //执行 GET 请求
                hResponse = httpclient.execute(httpGet);
                //连接成功
                if (hResponse.getStatusLine().getStatusCode() == 200)
{
                    //得到 HttpEntity 对象并转化为 String 类型
                    String strResult = EntityUtils.toString
                    (hResponse.getEntity());
                    geTextView = (TextView)findViewById
                    (R.id.textView1);
                    //显示 HttpEntity
                    geTextView.setText(strResult);
                }
            } catch (ClientProtocolException e) {
                // TODO Auto-generated catch block
                e.printStackTrace();
            } catch (IOException e) {
```

```
                              // TODO Auto-generated catch block
                              e.printStackTrace();
                       }
                }
         });
   }
}
```

在 AndroidMenifest.xml 中添加用户权限,允许应用程序访问网络。

```
<uses-permission android:name="android.permission.INTERNET" />
```

运行程序,单击 GET 按钮,效果如图 9.8 所示。

【示例 9-5】 下面通过案例演示,如何使用 POST 方法通信。

我们先来看这样一个例子。

(1)在浏览器地址栏中输入 http://192.168.1.102/zhishidian/test/login.asp,按回车键,打开对应页面,如图 9.9 所示。

图 9.8 HttpClient 的 GET 请求示例图   图 9.9 登录界面

(2)输入用户名和密码,单击"登录"按钮,进入 http://192.168.1.102/zhishidian/test/ogincheck.asp 页面,如图 9.10 所示。

下面我们开发一个 Android 程序,使用 HTTP 的 POST 方法实现 Android 客户端与网络之间的通信。在 HttpClient 项目的主界面添加一个 POST 按钮,用于发送 POST 请求。再添加一个空白文本框,用于显示网络数据信息。

逻辑代码如下:

```
post = (Button)findViewById(R.id.button2);
    post.setOnClickListener(new OnClickListener() {
         public void onClick(View v) {
              // TODO Auto-generated method stub
```

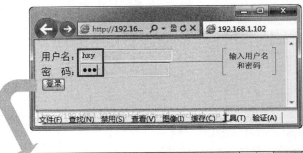

图9.10 登录成功

```
//HttpPost 连接对象
HttpPost hPost = new HttpPost("http://192.168.1.102/
zhishidian/test/logincheck.asp");
//使用 NameValuePair 来保存要传递的 Post 参数
List<NameValuePair> params = new ArrayList
<NameValuePair>();
//添加要传递的参数
params.add(new BasicNameValuePair("username", "hxy"));
params.add(new BasicNameValuePair("pwd", "123"));
try {
    //设置字符集
    HttpEntity httpentity = new UrlEncodedFormEntity
    (params, "gb2312");
    //请求 HttpRequest
    hPost.setEntity(httpentity);
    //取得默认的 HttpClient
    HttpClient httpclient = new DefaultHttpClient();
    //声明 HttpResponse
    HttpResponse httpResponse;
    //获取 HttpResponse
    httpResponse = httpclient.execute(hPost);
    //连接成功
    if (httpResponse.getStatusLine().getStatusCode() == 200){
    //获取返回的信息
        String result = EntityUtils.toString
          (httpResponse.getEntity());
        String str = new String(result.getBytes
        ("ISO 8859 1"),"gbk") ;
    //显示返回的信息
        posTextView = (TextView)findViewById
        (R.id.textView2);
        posTextView.setText(str);
            }
} catch (Exception e) {
    // TODO Auto-generated catch block
    e.printStackTrace();
}
```

        }
    });
```

运行程序，单击 POST 按钮，效果如图 9.11 所示。

图 9.11　HttpClient 的 POST 请求示例图

## 9.3　URL 网络通信

URL（Uniform Resource Locator）统一资源定位符，表示 Intetnet 上的资源地址，实现对网络资源的定位。简单地说，URL 就是 Web 地址，俗称"网址"。

### 9.3.1　URL 简介

URL 支持 HTTP、File、FTP 等多种协议。Java 通过 URL 标识，可以直接使用 HTTP、File、FTP 等多种协议，获取远端计算机上的资源信息，方便快捷地开发 Internet 应用程序。

#### 1. 统一资源定位器 URL

URL 的语法格式是<传输协议名>://<主机名>:<端口号>/<文件名>#<引用>。其中，<端口号>、<文件名>和<引用>是可选的，<传输协议名>和<主机名>是必需的。当没有给出<传输协议名>时，浏览器默认的传输协议是 HTTP。下面都是合法的 URL：

- http://www.sun.com；
- http://172.17.99.3；
- http://localhost:80；
- http://home.netscape.com/home/welcome.html；
- http://www.china.com/index.html#a。

#### 2. URL 类

java.net 包中定义了 URL 类。URL 类表示一个统一资源定位器。它是指向互联网上某
```

一资源的指针，这个资源可以是某个主机的一个文件或路径，也可以是文件上的一个锚（或称引用）。URL 构造方法如表 9-5 所示。

表 9-5 URL 构造方法

方 法 名 称	方 法 说 明
URL(String spec)	解析 spec，创建 URL
URL(URL context, String spec)	通过上下文，对 spec 解析创建 URL
URL(String protocol, String host, String file)	根据指定的 protocol、host 和 file，创建 URL
URL(String protocol, String host, int port, String file)	根据指定的 protocol、host、port 和 file，创建 URL

得到 URL 对象之后，就可以调用它的相关方法，获取相关信息。常用方法如表 9-6 所示。

表 9-6 URL 常用方法

方 法 名 称	方 法 说 明
String getProtocol()	返回当前 URL 的协议名
String getHost()	返回当前 URL 的主机名
int getPort()	返回当前 URL 的端口号
String getFile()	返回当前 URL 的文件名
String getQuery()	返回当前 URL 的查询
String getPath()	返回当前 URL 的路径
String getAuthority()	返回当前 URL 的权限
String getUserInfo()	返回当前 URL 的用户信息
String getRef()	返回当前 URL 的引用
InputStream openStream()	打开当前 URL 的连接，返回从这个连接读取的输入流
URLConnection openConnection()	返回一个由 URL 指示的表示与远程对象连接的 URLConnection 对象

9.3.2 URL 通信开发

URL 通信开发，分为以下几个步骤：

（1）根据指定的 URL 网址，创建 URL 对象；

```
URL myUrl = new URL(String spec);
```

（2）调用 URLConnection.openConnection()方法打开连接；

```
URLConnection uCoon = myUrl.openConnection();
```

（3）获取输入流；

```
InputStream in = uCoon.getInputStream();
```

（4）将网络信息提取显示。

【示例 9-6】 使用 URL 通信获取网络图片资源。新建项目 URL。在界面添加 Click 按钮，单击按钮获取网络图片。添加一个 ImageView 控件，用于显示获取到的网络图片。

逻辑代码如下：

```java
@TargetApi(11)
public class URLActivity extends Activity {
    private ImageView image;
    private Button click;
    @Override
    public void onCreate(Bundle savedInstanceState) {
        super.onCreate(savedInstanceState);
        setContentView(R.layout.activity_url);
    //添加以下两段代码,防止程序抛出NetworkOnMainThreadException异常
    StrictMode.setThreadPolicy(new StrictMode.ThreadPolicy.Builder()
    .detectDiskWrites()
    .detectDiskReads()
    .detectNetwork()
    .penaltyLog()
    .build());

    StrictMode.setVmPolicy(new StrictMode.VmPolicy.Builder()
    .detectLeakedSqlLiteObjects()
    .detectLeakedClosableObjects()
    .penaltyLog()
    .penaltyDeath()
    .build());
        click = (Button)findViewById(R.id.button1);
        click.setOnClickListener(new OnClickListener() {

            public void onClick(View v) {
                // TODO Auto-generated method stub
                try {
                    //根据字符串创建Url
                    URL myUrl = new URL("http://www.baidu.com/img/
                    baidu_sylogo1.gif");
                    //打开链接
                    URLConnection uCoon = myUrl.openConnection();
                    //获取输入流
                    InputStream in = uCoon.getInputStream();
                    //创建Bitmap
                    Bitmap bitmap = BitmapFactory.decodeStream(in);
                    //显示网络图片
                    image = (ImageView)findViewById(R.id.imageView1);
                    image.setImageBitmap(bitmap);
                } catch (IOException e) {
                    // TODO Auto-generated catch block
                    e.printStackTrace();
                }
            }
        });
    }
}
```

运行程序，单击 Click 按钮，效果如图 9.12 所示。

图 9.12 URL 通信示例图

9.4 WebView 网页开发

Android 系统提供内置的高性能浏览器，该浏览器应用了开源框架的 WebKit，在其 SDK 中封装了一个叫做 WebView 的控件。WebKit 浏览器的作用如图 9.13 所示，将 HTML 代码解释编译成直观、具体、用户可理解的网页界面。

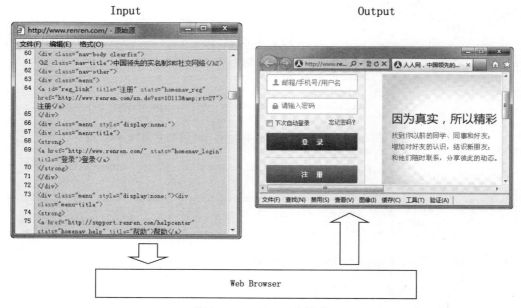

图 9.13 Webkit 工作模式

9.4.1 WebView 简介

WebView 是用于加载显示网页的控件。使用时，可以在布局界面中直接添加，然后在逻辑代码部分通过 ID 获取引用。WebView 的常用方法如表 9-7 所列。

表 9-7　WebView 常用方法

方法名称	方法说明
canGoBack()	得到 WebView 是否有一个向后的历史记录
canGoForward()	得到 WebView 是否有一个向前的历史记录
goBack()	在 WebView 历史记录中后退
goForward()	在 WebView 历史记录中前进
loadData(String data, String mimeType, String encoding)	加载执行给定数据的 URL
loadUrl(String url)	执行指定的 URL
SetWebChromeClient(WebChromeClient client)	设置 WebChromeClient
setWebViewClient(WebViewClient client)	设置 WebViewClient，接收各种通知和请求

9.4.2　WebView 开发

在 WebView 中，不仅可以根据指定的 URL 浏览网页，还可以载入 HTML 标记并显示。下面通过具体案例来说明。

【示例 9-7】　使用 WebView 浏览网页。新建项目 WebView。在界面添加 BACK 按钮，用于返回到上一历史记录；添加 NEXT 按钮用于前进到下一历史记录；添加 GO 按钮用于加载指定的 URL 网页；添加编辑框用于输入 URL 网址。添加 WebView 控件，用于浏览网页。程序界面如图 9.14 所示。

控件	属性	值
EditText	id	@+id/editText1
	inputType	textPostalAddress
Button	id	@+id/button1
	text	BACK
Button	id	@+id/button2
	text	NEXT
Button	id	@+id/button3
	text	GO
WebView	id	@+id/webView1

图 9.14　WebView 界面图

逻辑代码如下：

```
public class WebViewActivity extends Activity {
    private Button back,next,go;
    private EditText eText;
    private WebView webView;
    @Override
    public void onCreate(Bundle savedInstanceState) {
        super.onCreate(savedInstanceState);
        setContentView(R.layout.activity_web_view);
```

```java
eText = (EditText)findViewById(R.id.editText1);
webView = (WebView)findViewById(R.id.webView1);
//打开网页
go = (Button)findViewById(R.id.button3);
go.setOnClickListener(new OnClickListener() {

    public void onClick(View v) {
        // TODO Auto-generated method stub
        //获取编辑框中的 URL 网址
        String url = eText.getText().toString().trim();
        //确定 URL 为正确的网址
        if (URLUtil.isNetworkUrl(url)) {
        // WebView 打开指定的网址
            webView.loadUrl(url);
        }else {
            Toast.makeText(WebViewActivity.this, "网址有误",
                Toast.LENGTH_LONG).show();
        }
    }
});
//返回上一历史记录
back = (Button)findViewById(R.id.button1);
back.setOnClickListener(new OnClickListener() {

    public void onClick(View v) {
        // TODO Auto-generated method stub
        // WebView 有向前的历史记录
        if (webView.canGoBack()) {
        //返回上一历史记录
            webView.goBack();
        }else{
            Toast.makeText(WebViewActivity.this, "对不起,您现在不能后退",
                Toast.LENGTH_LONG).show();
        }
    }
});
//前进到下一历史记录
next = (Button)findViewById(R.id.button2);
next.setOnClickListener(new OnClickListener() {
    public void onClick(View v) {
        // TODO Auto-generated method stub
        // WebView 有向后的历史记录
        if (webView.canGoForward()) {
            //前进到下一历史记录
            webView.goForward();
        }else {
            Toast.makeText(WebViewActivity.this, "对不起,您现在不能前进",
                Toast.LENGTH_LONG).show();
        }
    }
});
}
}
```

在 AndroidMenifest.xml 中添加用户权限,允许应用程序访问网络。

```xml
<uses-permission android:name="android.permission.INTERNET" />
```

运行程序,效果如图 9.15 所示。

图 9.15 WebView 浏览网页

【示例 9-8】 使用 WebView 执行 HTML 代码,并显示对应网页。新建项目 HTML。在界面添加一个编辑框,用于输入 HTML 代码;添加一个 Click 按钮,单击时执行 HTML;添加一个 WebView 控件,用于显示 HTML 对应的网页。程序界面如图 9.16 所示。

控件	属 性	值
EditText	id	@+id/editText1
	inputType	textPostalAddress
Button	id	@+id/button1
	text	Click
WebView	id	@+id/webView1

图 9.16 HTML 界面图

逻辑代码如下:

```
public class HTMLActivity extends Activity {
    private EditText html;
    private Button click;
    private WebView webView;
    @Override
    public void onCreate(Bundle savedInstanceState) {
        super.onCreate(savedInstanceState);
        setContentView(R.layout.activity_html);

        click = (Button)findViewById(R.id.button1);
        click.setOnClickListener(new OnClickListener() {

            public void onClick(View v) {
                // TODO Auto-generated method stub
```

```
            html = (EditText)findViewById(R.id.editText1);
            //获取 HTML 代码
            String data = html.getText().toString().trim();
            webView = (WebView)findViewById(R.id.webView1);
            //执行 HTML 代码
            webView.loadData(data, "text/html", HTTP.UTF_8);
        }
    });
    }
}
```

在 AndroidMenifest.xml 中添加用户权限，允许应用程序访问网络。

```
<uses-permission android:name="android.permission.INTERNET" />
```

运行程序，在编辑框中输入"<html><head></head><body><a href=http://www.google.com>Click Here</a></body></html>"。单击 Click 按钮，在 WebView 中显示 HTML 代码内容 Click Here。单击 Click Here，界面跳转到谷歌首页，效果如图 9.17 所示。

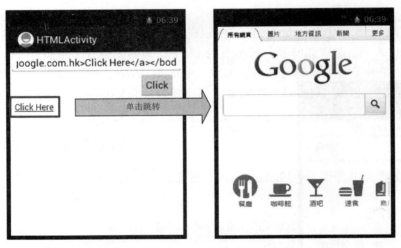

图 9.17　WebView 执行 HTML 代码

9.5　小　　结

本章内容主要介绍了 Android 系统中的通信方式。其中，Socket 通信、URL 通信较为简单；HTTP 通信和 WebView 网络开发是本章难点，需要读者多多练习，熟练掌握。在开发过程中，要保持计算机网络可用，切记添加"<uses-permission android:name="android.permission.INTERNET"/>"用户权限，允许程序访问网络。

9.6　习　　题

新建项目 WebView，在布局文件中添加 WebView 控件。在编辑框内输入网址，单击 GO 按钮，使用 WebView 浏览网页，如图 9.18 所示。

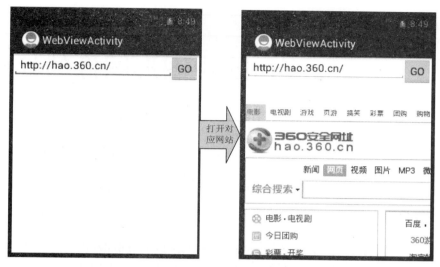

图 9.18　使用 WebView 浏览网页

【分析】本题目考查了读者对 WebView 的掌握。在开发时，要注意在 AndroidManifest.xml 文件中添加用户权限 "android.permission.INTERNET"，允许程序访问网络。可以参考 9.4.1 节的内容。

【核心代码】本题的核心代码如下所示。

```java
go.setOnClickListener(new OnClickListener() {
    public void onClick(View v) {
        // TODO Auto-generated method stub
        String url = eText.getText().toString().trim();
        if (URLUtil.isNetworkUrl(url)) {
            webView.loadUrl(url);
        }else {
            Toast.makeText(WebViewActivity.this, "网址有误",
                Toast.LENGTH_LONG).show();
        }
    }
});
```

第 10 章 Android 中图形图像的处理

图形图像的应用在一个系统中占有比较大的分量,如一些程序的图标、界面的美化等,都离不开图形图像。Android 中对图形图像的处理非常强大,对于 2D 图像它并没有沿用 Java 中的图形处理类,而是使用了自定义的处理类。本章将讲解 Bitmap 位图的使用、动画的创建和 Canvas、Paint 的基本绘图。

10.1 Android 中图形图像资源的获取

在之前的应用程序中,我们使用的几乎都是存储在 drawable 文件夹中的图片资源。本节将介绍一种新的提供图片的路径,即从 assets 文件夹中获取图片资源。例如我们如果想要将 SD 卡中的图片作为手机墙纸,就需要使用 Bitmap 和 BitmapFactory 类。

10.1.1 Bitmap 和 BitmapFactory 类

Bitmap 代表一张位图,BitmapDrawable 里封装的图片就是一个 Bitmap 对象。如果需要获取 BitmapDrawable 所包装的 Bitmap 对象,需要调用 BitmapDrawable.getBitmap()方法。语法如下:

```
Bitmap bitmap = bDrawable.getBitmap();
                    // 使用 getBitmap()方法来获取 BitmapDrawable 中的 Bitmap
```

Bitmap 相关方法如表 10-1 所示。

表 10-1 Bitmap 相关方法

方 法 名 称	方 法 说 明
public static BitmapcreateBitmap(Bitmap src)	返回一个 Bitmap 位图
public final boolean isRecycled()	判断该 Bitmap 对象是否被回收
public void recycle()	强制回收该 Bitmap 对象

BitmapFactory 是一个工具类,它提供了大量的方法,用于从不同的数据源解析、创建 Bitmap 对象。相关方法如表 10-2 所示。

表 10-2 BitmapFactory 相关方法

方 法 名 称	方 法 说 明
public static BitmapdecodeByteArray(byte[] data, int offset, int length)	从指定字节数组的 offset 位置开始,解析长度为 length 的字节数据为 Bitmap 对象
public static BitmapdecodeFile(String pathName)	从 pathName 指定的文件中解析创建 Bitmap 对象
public static BitmapdecodeResource(Resources res, int id)	根据 ID 指定的资源解析创建 Bitmap 对象
public static BitmapdecodeStream(InputStream is)	根从指定的输入流中解析创建 Bitmap 对象

10.1.2 获取 assets 文件夹图片资源

assets 文件夹里面的文件都是保持原始的文件格式，需要用 AssetManager 以字节流的形式读取文件。相关方法如表 10-3 所示。

表 10-3 AssetManager 相关方法

方 法 名 称	方 法 说 明
public void close()	关闭 AssetManager
public final InputStreamopen(String fileName)	打开指定资源对应的输入流
public final String[]list(String path)	返回指定路径下的所有文件

访问 assets 文件夹中的文件，分为以下几个步骤。

（1）在 Activity 里面调用 getAssets()方法，获取 AssetManager 引用。

```
private AssetManager assetManager;          // 声明 AssetManager 对象
assetManager = getAssets();                 // 获取 AssetManager 引用
```

（2）调用 AssetManager.open(String fileName)方法，指定读取的文件，得到输入流 InputStream。

```
inputStream = assetManager.open(String fileName);
                                            // 打开指定资源对应的输入流
```

（3）用已经 open ()方法建立的 inputStream 读取文件，读取完成后调用 inputStream.close()关闭输入流。

（4）调用 AssetManager.close()关闭 AssetManager。

注意：来自 assets 中的文件只可以读取，不能进行写的操作。即我们只能使用 assets 中的资源，不能更改。

【示例 10-1】 访问 assets 文件夹中的图片文件。新建项目 Bitmap。在 assets 文件夹下新建 logo 文件夹，保存一组图片在该文件夹中。在布局文件中添加一个 ImageView 控件用于显示图片；再添加一个按钮，单击按钮显示下一张图片。

逻辑代码如下：

```java
public class BitmapActivity extends Activity {
    private ImageView imageView;
    private Button btnNext;
    //存放图片资源的数组
    private String[] files;
    //声明 AssetManagerdx
    private AssetManager assetManager;
    //声明 Bitmap 对象
    private Bitmap bitmap;
    //数组下标
    private int currentImage = 0;
    /** Called when the activity is first created. */
    @Override
    public void onCreate(Bundle savedInstanceState) {
        super.onCreate(savedInstanceState);
        setContentView(R.layout.activity_bitmap);
```

```java
        imageView = (ImageView)findViewById(R.id.imageView1);
        btnNext = (Button)findViewById(R.id.button1);
        //获取 AssetManager 引用
        assetManager = getAssets();
        try {
        //返回 logo 文件夹下所有图片
         files = assetManager.list("logo");
         for (int i = 0; i < files.length; i++) {
         System.out.println(files[i]);
            }
        } catch (IOException e) {
            // TODO Auto-generated catch block
            e.printStackTrace();
        }
        btnNext.setOnClickListener(new OnClickListener() {
            public void onClick(View v) {
                if (currentImage >= files.length) {
                    currentImage = 0;
                }
                //判断是否为图片格式
                while (!files[currentImage].endsWith(".png")
                        &&!files[currentImage].endsWith(".jpg")
                        &&!files[currentImage].endsWith(".gif")) {
                    //显示下一张图片
                    currentImage++;
                //如果数组越界显示第一张图片
                if (currentImage >= files.length) {
                    currentImage = 0;
                }

                }

                InputStream inputStream = null;
                try {
                    //打开对应的输入流
                    inputStream = assetManager.open("logo/"+
                    files[currentImage++]);
                } catch (IOException e) {
                    // TODO Auto-generated catch block
                    e.printStackTrace();
                }
                //在界面中显示 Bitmap
                BitmapDrawable bDrawable = (BitmapDrawable)imageView.
                getDrawable();
                //强制回收
                if (bDrawable != null) {
                    if (!bDrawable.getBitmap().isRecycled()) {
                        bDrawable.getBitmap().recycle();
                    }
                    //改变 ImageView 显示的图片
                    bitmap = BitmapFactory.decodeStream(inputStream);
                    imageView.setImageBitmap(bitmap);
                }
            }
        });
    }
}
```

运行程序，效果如图 10.1 所示。

图 10.1　Bitmap 位图

10.2　Android 中的动画生成

Android 系统提供了两种创建动画的方式：补间动画（Tween Animation）和帧动画（Frame Animation）。补间动画（Tween Animation）主要实现对图片进行移动、放大、缩小，以及透明度变化的功能，而帧动画（Frame Animation）比较简单，就是将一张张的图片连续播放以产生动画效果。下面将分别介绍这两种动画技术的开发及应用。

10.2.1　补间动画

补间动画（Tween Animation）就是对场景里的对象不断地进行图像变化来产生动画效果，可以对对象进行旋转、平移、放缩和渐变等操作。表 10-4 列出了补间动画的几种变换的标记及属性值说明。

表 10-4　Tween Animation 中标签及属性值说明

标记名称	属　性　值	说　　明
\<set\>	shareInterpolator：是否在子元素中共享插入器	可以包含其他动画变换的容器，同时也可以包含\<set\>标记
\<alpha\>	fromAlpha：变换的起始透明度 toAlpha：变换的终止透明度，取值为 0.0~1.0，其中 0.0 代表全透明	实现透明度变换效果
\<scale\>	fromXScale：起始的 X 方向上的尺寸 toXScale：终止的 X 方向上的尺寸 fromYScale：起始的 Y 方向上的尺寸 toYScale：终止的 Y 方向上的尺寸；其中 1.0 代表原始大小 pivotX：进行尺寸变换的中心 X 坐标 pivotY：进行尺寸变换的中心 Y 坐标	实现尺寸变换效果，可以指定一个变换中心，例如指定 pivotX 和 pivotY 为(0,0)，则尺寸的拉伸或收缩均从左上角的位置开始

标记名称	属性值	说明
<translate>	fromXDelta：起始 X 位置 toXDelta：终止 Y 位置 fromYDelta：起始 Y 位置 toYDelta：终止 Y 位置	实现水平或竖直方向上的移动效果。如果属性值以"%"结尾，代表相对于自身的比例；如果以"%p"结尾，代表相对于父控件的比例；如果不以任何后缀结尾，代表绝对的值
<rotate>	fromDegree：开始旋转位置 toDegree：结束旋转位置；以角度为单位 pivotX：旋转中心点的 X 坐标 pivotY：旋转中心点的 Y 坐标	实现旋转效果，可以指定旋转定位点

表 10-4 列出了各个标签中特有的属性，下面介绍 XML 文件中标签的一些共有属性，如表 10-5 所示。

表 10-5　Tween Animation 中标签共有属性值说明

属性值	说明
duration	变换持续的时间，以毫秒为单位
startOffset	变换开始的时间，以毫秒为单位
repeatCount	定义该动画重复的次数
interpolator	为每个子标记变换设置插入器，系统已经设置好一些插入器，可以在 R.anim 包下找到

【示例 10-2】 补间动画的使用。新建项目 Tween，在程序中添加一个 ImageView 显示动画图片，添加一个 Button，单击开始播放动画，并拷贝程序中会用到的图片资源到 res/drawable 目录下。在 Eclipse 中开发一个定义了补间动画的 XML 文件 tween.xml，位于 res/anim 目录下（该目录需要手动创建），添加代码产生透明度变化效果。然后在逻辑代码部分调用方法加载动画图片并播放。

tween.xml：

```xml
<?xml version="1.0" encoding="UTF-8"?>
<set xmlns:android="http://schemas.android.com/apk/res/android">
    <!--透明度变化   -->
    <alpha

        android:fromAlpha="1.0" <!--透明度初始值   -->
        android:toAlpha="0.0"<!--透明度最终值   -->
        android:duration="10000"<!--变化时长   -->

    />
</set>
```

逻辑代码：

```java
public class TweenActivity extends Activity {
    ImageView imageView;
    Button button;
    @Override
    public void onCreate(Bundle savedInstanceState) {
        super.onCreate(savedInstanceState);
```

```
setContentView(R.layout.activity_tween);
imageView = (ImageView)findViewById(R.id.imageView1);
button = (Button)findViewById(R.id.button1);
button.setOnClickListener(new OnClickListener() {
    public void onClick(View v) {
        // TODO Auto-generated method stub
        //加载动画图片
        Animation animation = AnimationUtils.loadAnimation
          (TweenActivity.this, R.anim.tween);
        //播放动画
        imageView.startAnimation(animation);
    }
});
}
```

运行程序，效果如图 10.2 所示。

初始状态
0毫秒

逐渐透明
8000毫秒左右

完全透明
10000毫秒

图 10.2　图片透明度变化

【示例 10-3】　修改【示例 10-2】中的 tween.xml 动画文件，实现图片尺寸大小变化的效果。运行程序，效果如图 10.3 所示。

tween.xml：

```
<?xml version="1.0" encoding="UTF-8"?>
<set xmlns:android="http://schemas.android.com/apk/res/android">
    <!--尺寸变化  -->
    <scale
            <!--起始时 X 方向上的尺寸-->
            android:fromXScale="1.0"
            <!--终止时 X 方向上的尺寸-->
            android:toXScale="0.0"
            <!--起始时 Y 方向上的尺寸-->
            android:fromYScale="1.0"
            <!--终止时 Y 方向上的尺寸-->
            android:toYScale="0.0"
            <!--尺寸变换初中心 X 坐标-->
            android:pivotX="50%"
            <!--尺寸变换初中心 Y 坐标-->
            android:pivotY="50%"
            <!--变换时长-->
            android:duration="10000"
    />
```

```
</set>
```

原始大小　　　　　　　　逐渐缩小　　　　　　　　完全消失
0毫秒　　　　　　　　6000毫秒左右　　　　　　10000毫秒

图 10.3　图片尺寸大小变化

【示例 10-4】 修改【示例 10-2】中的 tween.xml 动画文件，实现图片位置变化的效果。运行程序，效果如图 10.4 所示。

tween.xml：

```xml
<?xml version="1.0" encoding="UTF-8"?>
<set xmlns:android="http://schemas.android.com/apk/res/android">
    <!--位置变化  -->
     <translate
            <!--起始 X 位置-->
            android:fromXDelta="30"
            <!--终止 X 位置-->
            android:toXDelta="0"
            <!--起始 Y 位置-->
            android:fromYDelta="30"
            <!--终止 Y 位置-->
            android:toYDelta="0"
            <!--变换时长-->
            android:duration="10000"
     />
</set>
```

初始位置　　　　　　　　位置变化　　　　　　　　最终位置
0毫秒　　　　　　　　6000毫秒左右　　　　　　10000毫秒

图 10.4　图片位置变化

【示例 10-5】 修改【示例 10-2】中的 tween.xml 动画文件，实现图片旋转的效果。运行程序，效果如图 10.5 所示。

tween.xml：

```xml
<?xml version="1.0" encoding="UTF-8"?>
<set xmlns:android="http://schemas.android.com/apk/res/android">
    <!--旋转变化   -->
    <rotate
            <!--初始角度-->
            android:fromDegrees="0"
            <!--初始角度-->
            android:toDegrees="+360"
            <!--旋转中心点的 X 坐标-->
            android:pivotX="50%"
            <!--旋转中心点的 Y 坐标-->
            android:pivotY="50%"
            <!--旋转时长-->
            android:duration="10000"
            />
</set>
```

初始位置

90°旋转

180°旋转

图 10.5 图片旋转

10.2.2 帧动画

帧动画（Frame Animation）就如同电影一样，通过顺序播放一系列事先加载好的静态图片产生动画效果。帧动画的 XML 文件中主要是用到的标签及其属性，如表 10-6 所示。

表 10-6 Frame Animation 中标签及其属性说明

标签名称	属 性 值	说　　明
\<animation-list\>	android:oneshot：如果设置为 true，则该动画只播放一次，然后停止在最后一帧	Frame Animation 的根标记，包含若干\<item\>标记
\<item\>	android:drawable：图片帧的引用 android:duration：图片帧的停留时间 android:visible：图片帧是否可见	每个\<item\>标记定义了一个图片帧，其中包含图片资源的引用等属性

【示例 10-6】 帧动画的实现。新建项目 Frame，在程序中添加一个 ImageView 显示动画图片，添加一个 Button，单击开始播放动画，并拷贝程序中会用到的图片资源到 res/drawable 目录下。在 Eclipse 中开发一个定义了帧动画的 XML 文件，本项目为 frame.xml，位于 res/anim 目录下（该目录需要手动创建）。然后在逻辑代码部分调用方法加载动画图片并播放。

frame.xml：

```xml
<?xml version="1.0" encoding="UTF-8"?>
<animation-list
    xmlns:android="http://schemas.android.com/apk/res/android"
    <!--设置动画循环播放-->
    android:oneshot="false" >
    <!--添加 4 张图片每张显示 500 毫秒-->
 <item android:duration="500" android:drawable="@drawable/png1132"/>
 <item android:duration="500" android:drawable="@drawable/png1139"/>
 <item android:duration="500" android:drawable="@drawable/png1144"/>
 <item android:duration="500" android:drawable="@drawable/png1145"/>

</animation-list>
```

布局文件：

```xml
<RelativeLayout xmlns:android="http://schemas.android.com/apk/res/android"
    xmlns:tools="http://schemas.android.com/tools"
    android:layout_width="match_parent"
    android:layout_height="match_parent" >
    <ImageView
        android:id="@+id/imageView1"
        android:layout_width="wrap_content"
        android:layout_height="wrap_content"
        android:layout_alignParentTop="true"
        android:layout_centerHorizontal="true"
        android:layout_marginTop="86dp"
        android:src="@anim/frame" /><!--引用帧动画文件-->
    <Button
        android:id="@+id/button1"
        android:layout_width="wrap_content"
        android:layout_height="wrap_content"
        android:layout_alignParentTop="true"
        android:layout_marginTop="26dp"
        android:layout_toRightOf="@+id/imageView1"
        android:text="click" />
</RelativeLayout>
```

逻辑代码：

```java
public class FrameActivity extends Activity {
    private ImageView imageView;
    private Button button;
    //声明创建逐帧动画的对象
    private AnimationDrawable aDrawable;
    //声明动画对象
    private Animation animation;
    @Override
    public void onCreate(Bundle savedInstanceState) {
        super.onCreate(savedInstanceState);
```

```
setContentView(R.layout.activity_frame);
imageView = (ImageView)findViewById(R.id.imageView1);
button = (Button)findViewById(R.id.button1);
//获取创建逐帧动画的实例对象
aDrawable =(AnimationDrawable)imageView.getDrawable();
button.setOnClickListener(new OnClickListener() {
    public void onClick(View v) {
        // TODO Auto-generated method stub
        //设置动画播放
        imageView.setAnimation(animation);
        //启动动画播放
        aDrawable.start();
    }
});
}
```

注意：在 Activity 的 onCreate()方法中调用 AnimationDrawable.start()方法，图片不会被播放。因为此时 AnimationDrawable 类尚未完全与 Window 类接触。我们需要添加一个触发事件，触发图片播放。

运行程序，效果如图 10.6 所示，顺序播放一系列图片。

图 10.6　帧动画

10.3　Android 中图形的绘制

在 Android 系统中，图形绘制功能也是很强大的，在程序开发中有一些控件需要自己去绘制，这时我们可以利用 Android 中的相应类去完成。特别是在游戏开发、界面设计中，很多图形的绘制都需要用户使用 Canvas（画布）类和 Paint（画笔）类进行绘图程序的开发。

10.3.1　图形绘制类介绍

在绘制一些图形时，需要用到 Android 中的一些类，如 Canvas 类、Paint 类等，它们在图形绘制过程中起到了至关重要的作用。下面将介绍这几个类的使用方法。

Canvas 类主要实现了屏幕的绘制过程，其中包含很多使用的方法，如绘制一条路径、区域、贴图、画点、画线、渲染文本等。Canvas 的绘图方法如表 10-7 所示。

表 10-7 Canvas 绘图方法

方 法 名 称	方法说明
public boolean clipRect(float left, float top, float right, float bottom)	剪切一个矩形区域
public boolean clipRegion(Region region)	剪切指定区域
public void drawArc(RectF oval, float startAngle, float sweepAngle, boolean useCenter, Paint paint)	画弧
public void drawCircle(float cx, float cy, float radius, Paint paint)	绘制圆形
public void drawLine(float startX, float startY, float stopX, float stopY, Paint paint)	绘制直线
public void drawOval(RectF oval, Paint paint)	绘制椭圆
public void drawRect(RectF rect, Paint paint)	绘制矩形
public void drawRoundRect(RectF rect, float rx, float ry, Paint paint)	绘制圆角矩形

在 Canvas 类中，有很多方法需要使用 Paint 类作为参数，由此可见 Paint 类的重要性，Paint 就是绘画的工具，如画笔、画刷、颜料等。它包含了很多方法，主要方法如表 10-8 所示。

表 10-8 Paint 相关方法

方 法 名 称	方法说明
public void setARGB(int a, int r, int g, int b)	设置颜色
public void setAlpha(int a)	设置透明度
public void setAntiAlias(boolean aa)	设置是否抗锯齿
public void setColor(int color)	设置颜色
public void setStrokeWidth(float width)	设置笔触宽度
public void setStyle(Paint.Style style)	设置填充风格

10.3.2 基本图形的绘制

学习了上面的相关 API，下面来实现在 Canvas 上绘制图形。

【示例 10-7】 Android 中基本图形的绘制。新建项目 Canvas。再新建类 CanvasView 继承于 View 类，添加其构造方法，并重写 onDraw()方法。在 onDraw()方法中，调用 Canvas 的绘图方法绘制各种基本图形。然后在 CanvasActivity 中加载 CanvasView 对象，显示图形。

CanvasView：

```java
//继承于View类
public class CanvasView extends View {
    //画笔
    private Paint paint;
    //构造方法
    public CanvasView(Context context) {
        super(context);
        // TODO Auto-generated constructor stub
        //创建画笔
        paint = new Paint();
    }
```

```java
//重写onDraw()方法
@Override
protected void onDraw(Canvas canvas) {
    // TODO Auto-generated method stub
    super.onDraw(canvas);
    //设置画布为白色
    canvas.drawColor(Color.WHITE);
    //设置画笔为蓝色
    paint.setColor(Color.BLUE);
    //设置画笔为实心
    paint.setStyle(Paint.Style. FILL);
    //设置画笔笔触宽度为5
    paint.setStrokeWidth(5);
    //绘制实心正方形
    canvas.drawRect(10,80,70,140, paint);
    //绘制实心圆形
    canvas.drawCircle(40, 40, 30, paint);
    //绘制实心矩形
    canvas.drawRect(10, 150, 70, 190, paint);
    //绘制实心圆角矩形
    RectF rectf1 = new RectF(10,200,70,230);
    canvas.drawRoundRect(rectf1, 15, 15, paint);
    //绘制实心椭圆
    RectF rectf2 = new RectF(10,240,70,270);
    canvas.drawOval(rectf2, paint);
    //设置画笔为空心
    paint.setStyle(Paint.Style. STROKE);
    //绘制空心正方形
    canvas.drawRect(90,80,150,140, paint);
    //绘制空心矩形
    canvas.drawRect(90, 150, 150, 190, paint);
    //绘制空心圆形
    canvas.drawCircle(120, 40, 30, paint);
    //绘制空心圆角矩形
    RectF rectf3 = new RectF(90,200,150,230);
    canvas.drawRoundRect(rectf3, 15,15,paint);
    //绘制空心椭圆
    RectF rectf4 = new RectF(90,240,150,270);
    canvas.drawOval(rectf4, paint);
    }
}
```

CanvasActivity：

```java
public class CanvasActivity extends Activity {
    @Override
    public void onCreate(Bundle savedInstanceState) {
        super.onCreate(savedInstanceState);
        //加载CanvasView对象,显示图形在界面上
        setContentView(new CanvasView(this));
    }
}
```

运行程序，效果如图10.7所示。

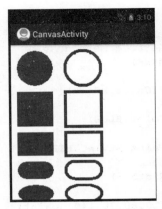

图 10.7　Canvas 效果图

10.4　小　　结

本章主要讲解了 Android 系统中图形的应用。其中，访问 assets 文件夹中的图片资源是本章难点，需要读者多多练习，熟练掌握。动画和 Canvas 绘图在游戏开发中比较常用，我们可以根据程序需要，引用或绘制各种各样的图形，丰富界面的多样性。

10.5　习　　题

新建项目 Tween，实现图片的位置、透明度、旋转和尺寸同时变化的动画，如图 10.8 所示。

图 10.8　Tween 动画

【分析】本题目主要考查读者对 Tween 动画的掌握，设置图片同时实现位置、透明度、旋转和尺寸的变化。需要在 res 目录下手动创建动画文件 tween.xml，可以参考 10.2.1 节的开发程序。

【核心代码】本题的核心代码如下所示。

tween.xml 文件：

```xml
<?xml version="1.0" encoding="UTF-8"?>
<set xmlns:android="http://schemas.android.com/apk/res/android">
    <rotate
        android:fromDegrees="0"
        android:toDegrees="+360"
        android:pivotX="50%"
        android:pivotY="50%"
        android:duration="10000"
        />
    <alpha
       android:fromAlpha="1.0"
       android:toAlpha="0.0"
       android:duration="10000"
       />
    <scale
        android:fromXScale="1.0"
         android:toXScale="0.0"
        android:fromYScale="1.0"
        android:toYScale="0.0"
        android:pivotX="50%"
        android:pivotY="50%"
        android:duration="10000"
         />
     <translate
            android:fromXDelta="30"
            android:toXDelta="0"
            android:fromYDelta="30"
            android:toYDelta="0"
            android:duration="10000"
        />
</set>
```

逻辑代码:

```
Animation animation = AnimationUtils.loadAnimation
                (TweenActivity.this, R.anim.tween);
imageView.startAnimation(animation);
```

第 11 章　Android 多媒体应用

每个使用 Android 系统的人都知道 Android 系统中带有一个图库应用程序和一个音乐播放器。打开图库可以查看当前终端里所有的图片文件,而音乐播放器可以看当前终端里所有的音乐文件。这就是 Android 的多媒体。除此之外,Android 多媒体还支持视频的播放和录制,以及图片的采集(即拍照)。

11.1　音乐播放器

音乐播放器是手机中的一个最基本的应用。在 Android 中,与音频相关的类是 MediaPlayer 类,它提供了音频的播放、暂停、停止和循环等功能方法。在 Android 系统中,支持的音频格式主要有 MP3、WAV 和 3GP。默认支持的音频文件有存储在应用程序中的本地资源(Resource)、存储在文件系统的标准音频文件(Local)、通过网络连接取得的数据流(URL)。

11.1.1　MediaPlayer 类简介

Android 系统使用 MediaPlayer 类来播放音频。MediaPlayer 类的相关方法,如表 11-1 所示。

表 11-1　MediaPlayer 类的相关方法

方 法 名 称	方 法 说 明
public static MediaPlayercreate(Context context, int resid)	从 resid 资源 ID 对应的资源文件中装载音频文件,并返回新创建的 MediaPlayer 对象
public static MediaPlayercreate(Context context, Uri uri)	从指定的 Uri 装载音频文件,并返回新创建的 MediaPlayer 对象
public int getDuration()	获取音频文件播放的总时长
public void pause()	暂停音乐播放
public void prepare()	准备播放器播放
public void reset()	重置未初始化状态的媒体播放器
public void seekTo(int msec)	寻求指定的时间位置,播放指定的音频内容
public void setDataSource(String path)	指定装载 path 路径所代表的文件
public void setDataSource(Context context, Uri uri)	指定装载 Uri 所代表的文件
public void setDataSource(FileDescriptor fd, long offset, long length)	指定装载 fd 所代表的文件中从 offset 开始,长度为 length 的文件内容
public void setDataSource(FileDescriptor fd)	指定装载 fd 所代表的文件

续表

方 法 名 称	方 法 说 明
public void setLooping(boolean looping)	设置循环播放
public void setVolume(float leftVolume, float rightVolume)	设置音乐音量
public void start()	播放音乐
public void stop()	停止播放音乐

11.1.2 本地音频文件播放

使用 MediaPlayer 播放存储在应用程序中的本地资源音频文件，流程如图 11.1 所示。

1. 音乐的播放、暂停和停止

【示例 11-1】 使用 MediaPlayer 播放存储在应用程序中的本地资源音频文件。新建项目 Resource。在界面添加 3 个按钮：Start 按钮用来播放音乐、Pause 按钮用来暂停音乐、Stop 按钮用来停止音乐播放。添加音频文件到 res/raw 目录下，该目录需要手动创建，程序界面如图 11.2 所示。

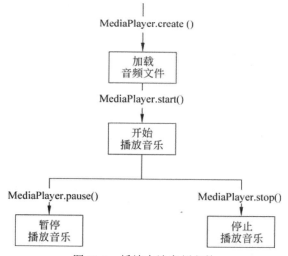

图 11.1 播放本地音频文件

图 11.2 程序界面图

逻辑代码如下：

```java
public class MediaPlayer1Activity extends Activity {
    //控制音乐播放、暂停和停止的按钮
    private Button start,pause,stop;
    //声明 MediaPlayer 类
    private MediaPlayer mPlayer;
    @Override
    public void onCreate(Bundle savedInstanceState) {
        super.onCreate(savedInstanceState);
        setContentView(R.layout.activity_media_player1);
        //创建 MediaPlayer 对象
        mPlayer = MediaPlayer.create(MediaPlayer1Activity.this, R.raw.destiny);
        start = (Button)findViewById(R.id.button1);
```

```java
start.setOnClickListener(new OnClickListener() {
    public void onClick(View v) {
        // TODO Auto-generated method stub
        //播放音乐
        mPlayer.start();
    }
});
pause = (Button)findViewById(R.id.button2);
pause.setOnClickListener(new OnClickListener() {
    public void onClick(View v) {
        // TODO Auto-generated method stub
        //暂停音乐
        mPlayer.pause();
    }
});
stop = (Button)findViewById(R.id.button3);
stop.setOnClickListener(new OnClickListener() {
    public void onClick(View v) {
        // TODO Auto-generated method stub
        //停止音乐
        mPlayer.stop();
    }
});
}
```

运行程序，单击 Start 按钮，音乐开始播放；单击 Pause 按钮，音乐暂停播放；单击 Stop 按钮，音乐停止播放。

2. 音量设置

Android 提供了 public void setVolume(float leftVolume, float rightVolume)方法设置音量，语法如下：

```
mPlayer.setVolume(float leftVolume,float rightVolume)
```

其中，leftVolume 表示左声道声音，rightVolume 表示右声道声音。leftVolume 和 rightVolume 参数值范围均为 0.0f（声音最小）～1.0f（声音最大）。

【示例 11-2】 修改【示例 11-1】，在 MediaPlayer1 界面再添加两个按钮：Up 按钮将声音置为最大、Down 按钮将声音置为最小，程序界面如图 11.3 所示。

逻辑代码如下：

```java
up = (Button)findViewById(R.id.button4);
up.setOnClickListener(new OnClickListener() {

    public void onClick(View v) {
        // TODO Auto-generated method stub
        //将声音置为最大
        mPlayer.setVolume(1.0f, 1.0f);
    }
});
down = (Button)findViewById(R.id.button5);
down.setOnClickListener(new OnClickListener() {
    public void onClick(View v) {
```

```
        // TODO Auto-generated method stub
        //将声音置为最小
        mPlayer.setVolume(0.0f, 0.0f);
    }
});
```

运行程序，单击 Up 按钮，音乐的音量变为最大；单击 Down 按钮，音乐的音量变为最小，即没有声音。

图 11.3 程序界面图

图 11.4 程序界面图

3. 播放进度设置

Android 提供了 public void seekTo(int msec)方法设置音乐的播放进度，播放指定的音频文件内容。语法如下：

```
mPlayer.seekTo(int msec);
```

其中，seekTo()方法用来查找指定时间的位置，msec 属性用来设定音频文件偏移时长。

【示例 11-3】 修改【示例 11-2】，在 MediaPlayer1 界面再添加一个 SeekBar 控件。拖动 SeekBar，调节音乐的播放进度，程序界面如图 11.4 所示。

逻辑代码如下：

```
seekBar.setOnSeekBarChangeListener(new OnSeekBarChangeListener() {
        //监听 seekbar 停止拖动事件
        public void onStopTrackingTouch(SeekBar seekBar) {
            // TODO Auto-generated method stub
            //获取 seekbar 当前进度值
            int dest = seekBar.getProgress();
            //获取音乐总时长
            int mMax = mPlayer.getDuration();
            //获取 seekbar 最大进度值
            int sMax = seekBar.getMax();
            //播放与当前 seekbar,进度对应的音乐内容
            mPlayer.seekTo(mMax* dest/sMax);
        }
        public void onStartTrackingTouch(SeekBar seekBar) {
            // TODO Auto-generated method stub
        }
        public void onProgressChanged(SeekBar seekBar, int progress,
```

```
            boolean fromUser) {
        // TODO Auto-generated method stub
    }
});
```

运行程序，拖动 SeekBar，音乐随拖动条拖动进度播放。

4. 音乐循环播放

Android 系统提供了 public void setLooping(boolean looping)方法，传入参数 true 时，设置音乐循环播放。

【示例 11-4】 修改【示例 11-3】，在 MediaPlayer1 界面再添加一个 Loop 按钮。单击按钮，设置音乐循环播放，程序界面如图 11.5 所示。

图 11.5　程序界面图

逻辑代码如下：

```
loop.setOnClickListener(new OnClickListener() {
    public void onClick(View v) {
        // TODO Auto-generated method stub
        //设置音乐循环播放
        mPlayer.setLooping(true);
    }
});
```

运行程序，单击 Loop 按钮。音乐完整播放后，继续循环播放。

11.1.3　多个标准音频文件播放

上一节中，调用 public static MediaPlayercreate(Context context, int resid)方法播放音乐，使用方法非常简单，但每次调用都会返回新创建的 MediaPlayer 对象。如果程序需要使用 MediaPlayer 播放多个音频文件，使用 MediaPlayer 的静态 create()方法就不合适了。此时我们可以考虑，使用 public void setDataSource(String path)方法来装载指定的音频文件。

【示例 11-5】 使用 MediaPlayer 播放标准音频文件。新建项目 Local。在界面添加 4 个按钮：Start 按钮播放音乐、Pause 按钮暂停音乐、Stop 按钮停止音乐播放、Next 按钮播放下一首音乐。然后添加音频文件到 SDCard，添加过程如图 11.6 所示。程序界面如图 11.7 所示。

图 11.6 添加音频文件到 SDCard

图 11.7 程序界面图

逻辑代码如下：

```java
public class MediaPlayer2Activity extends Activity {
    private Button start,pause,stop,next;
    @Override
    public void onCreate(Bundle savedInstanceState) {
        super.onCreate(savedInstanceState);
        setContentView(R.layout.activity_media_player2);
        //创建 MediaPlayer 对象
        final MediaPlayer mPlayer = new MediaPlayer();
        start = (Button)findViewById(R.id.button1);
        start.setOnClickListener(new OnClickListener() {

            public void onClick(View v) {
                // TODO Auto-generated method stub
                try {
                    //重置未初始化状态的 Mediaplayer
                    mPlayer.reset();
                    //加载指定的音频文件
                    mPlayer.setDataSource("/mnt/sdcard/miracle.mp3");
                    //准备 Mediaplayer 播放
                    mPlayer.prepare();
                    //播放音乐
                    mPlayer.start();
                } catch (IllegalArgumentException e) {
                    // TODO Auto-generated catch block
                    e.printStackTrace();
                } catch (SecurityException e) {
                    // TODO Auto-generated catch block
                    e.printStackTrace();
                } catch (IllegalStateException e) {
                    // TODO Auto-generated catch block
                    e.printStackTrace();
                } catch (IOException e) {
                    // TODO Auto-generated catch block
                    e.printStackTrace();
                }
            }
        });
        pause = (Button)findViewById(R.id.button2);
        pause.setOnClickListener(new OnClickListener() {
            public void onClick(View v) {
                // TODO Auto-generated method stub
                //暂停音乐
                mPlayer.pause();
```

```java
            }
        });
        stop = (Button)findViewById(R.id.button3);
        stop.setOnClickListener(new OnClickListener() {
            public void onClick(View v) {
                // TODO Auto-generated method stub
                //停止播放音乐
                mPlayer.stop();
            }
        });
        next = (Button)findViewById(R.id.button4);
        next.setOnClickListener(new OnClickListener() {
            public void onClick(View v) {
                // TODO Auto-generated method stub
                mPlayer.reset();
                try {
                //加载音频文件
                    mPlayer.setDataSource("/mnt/sdcard/takeitoff.mp3");
                //准备 Mediaplayer 播放
                    mPlayer.prepare();
                //播放音乐
                    mPlayer.start();
                } catch (IllegalArgumentException e) {
                    // TODO Auto-generated catch block
                    e.printStackTrace();
                } catch (SecurityException e) {
                    // TODO Auto-generated catch block
                    e.printStackTrace();
                } catch (IllegalStateException e) {
                    // TODO Auto-generated catch block
                    e.printStackTrace();
                } catch (IOException e) {
                    // TODO Auto-generated catch block
                    e.printStackTrace();
                }
            }
        });
    }
}
```

运行程序。单击 Start 按钮，音乐开始播放；单击 Pause 按钮，音乐暂停播放；单击 Stop 按钮，音乐停止播放；单击 Next 按钮，播放下一首指定音乐。

11.2 视频播放器

上一节中我们学习了有关音频播放的知识，本节介绍 Android 系统中视频播放方面的应用。Android 系统支持的视频文件格式有 3GP、MP4。Android 系统所能播放的视频文件可以存储在 SDCard 或 Android 的系统文件内。

11.2.1 视频相关类简介

Android 系统提供了 VideoView 控件，用于在界面设计时显示视频文件。在使用 VideoView 播放视频文件时，需要使用到以下类。

1. VideoView 类

VideoView 提供了一系列方法,如表 11-2 所示。

表 11-2 VideoView 相关方法

方 法 名 称	方 法 说 明
public boolean canPause ()	判断是否能够暂停播放视频
public boolean canSeekBackward ()	判断是否能够倒退
public boolean canSeekForward()	判断是否能够快进
public int getCurrentPosition ()	获得当前的位置
public int getDuration ()	获得所播放视频的总时间
public boolean isPlaying()	判断是否正在播放视频
public void pause ()	暂停视频播放
public void seekTo (int msec)	设置播放位置
public void setMediaController (MediaController controller)	设置媒体控制器
public void setVideoPath(String path)	设置 path 路径所代表的视频文件
public void setVideoURI(Uri uri)	加载 Uri 对应的视频文件
public void start ()	开始播放视频文件
public void stopPlayback()	停止回放视频文件

2. MediaController 类

MediaController 是一个包含了媒体播放器(MediaPlayer)控件的视图。包含了一些典型的按钮,比如"播放(Play)"、"暂停(Pause)"、"倒带(Rewind)"、"快进(Fast Forward)"与进度滑动器(Progress Slider)。它管理媒体播放器(MediaPlayer)的状态以保持控件的同步。

这个媒体控制器将创建一个具有默认设置的控件,并把它放到一个窗口里漂浮在应用程序之上。如果这个窗口空闲 3 秒钟,那么它将消失,直到用户触摸这个视图时重现。

注意:当媒体控制器是在一个 XML 布局资源文件中创建时,像 show()和 hide()这些函数是无效的。

11.2.2 视频播放流程

VideoView 是一个位于 android.widget 包下的组件,我们使用这个组件来完成视频的播放。使用 VideoView 播放视频文件的流程,如图 11.8 所示。

【示例 11-6】 使用 VideoView 播放视频文件。新建项目 VideoView。在布局文件中添加 VideoView 控件,用于播放视频文件。添加视频文件 video.3gp 到 SDCard 中,可以参考音频文件的导入方法来添加。

逻辑代码如下:

```
public class VideoViewActivity extends Activity {
    //声明 VideoView 对象
    private VideoView videoView;
    @Override
    public void onCreate(Bundle savedInstanceState) {
        super.onCreate(savedInstanceState);
```

```
setContentView(R.layout.activity_video_view);
//获取VideoView控件引用
videoView = (VideoView)findViewById(R.id.videoView1);
//创建MediaController对象
MediaController mController = new MediaController(this);
//加载视频文件
videoView.setVideoPath("/mnt/sdcard/video.3gp");
//设置Mediacontroller
videoView.setMediaController(mController);
//设置Mediacontroller与MediaPlayer关联
mController.setMediaPlayer(videoView);
    }
}
```

运行程序。单击视频界面，浮现 MediaController 媒体控制器。单击播放按钮，开始播放视频；再次单击，视频暂停播放。如图 11.9 所示。

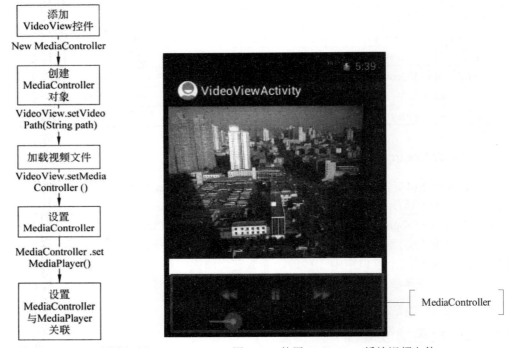

图 11.8　使用 VideoView 播放视频文件流程图　　　　图 11.9　使用 VideoView 播放视频文件

11.3　音频与视频的录制

Android 系统除了提供相关类，实现音频与视频的播放之外，还提供了 MediaRecorder 类实现音频和视频的录制，但是需要有硬件设备的支持。

11.3.1　音频录制

手机一般都提供了麦克风硬件，Android 系统可以利用该硬件录制音频。使用

MediaRecorder 类录制音频用到的相关方法，如表 11-3 所示。

表 11-3 MediaRecorder 相关方法

方 法 名 称	方 法 说 明
public MediaRecorder()	默认构造方法
public void prepare()	准备记录器开始捕捉和编码数据
public void release()	释放与此相关的 MediaRecorder 对象资源
public void setAudioEncoder(int audio_encoder)	设置音频编码格式
public void setAudioSource(int audio_source)	设置声音来源
public void setOutputFile(String path)	设置音频文件保存位置
public void setOutputFormat(int output_format)	设置所录制的音频文件的格式
public void start()	开始录制
public void stop()	结束录制

使用 MediaRecorder 类录制音频的开发流程如图 11.10 所示。

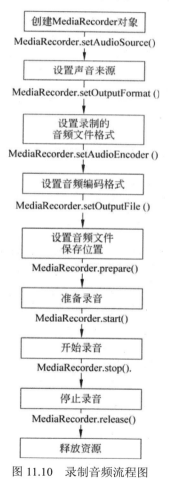

图 11.10 录制音频流程图

注意：在开发过程中，要先设置音频文件格式，再设置音频编码格式，否则程序抛出

IllegalStateException 异常。

【示例 11-7】 录制音频。新建项目 RecordSound。在界面添加"开始录音"按钮和"停止录音"按钮。录音成功后，在 mnt/sdcard 路径下生成 sound.amr 音频文件。

逻辑代码如下：

```java
public class RecordSoundActivity extends Activity {
    //开始录音和停止录音按钮
    private Button start,stop;
    //音频文件
    private File soundFile;
    //记录器
    private MediaRecorder recorder;
    @Override
    public void onCreate(Bundle savedInstanceState) {
        super.onCreate(savedInstanceState);
        setContentView(R.layout.activity_record_sound);

        start = (Button)findViewById(R.id.button1);
        start.setOnClickListener(new OnClickListener() {

            public void onClick(View v) {
                // TODO Auto-generated method stub
                //如果无 SDCard 可用,则使用 Toast 提示
                if (!Environment.getExternalStorageState().equals
                 (Environment.MEDIA_MOUNTED)) {
                    Toast.makeText(RecordSoundActivity.this, "请插入SDCard! ",
                        Toast.LENGTH_SHORT).show();
                    return;
                }
                try {
                    //创建保存录音的音频文件
                    soundFile = new
                    File(Environment.getExternalStorageDirectory().
                    getCanonicalFile() + "/sound.amr");
                    //创建记录器
                    recorder = new MediaRecorder();
                    //声音来源于麦克风
                    recorder.setAudioSource(MediaRecorder.
                    AudioSource.MIC);
                    //录制的音频文件,格式为 3GP
                    recorder.setOutputFormat(MediaRecorder.
                    OutputFormat.THREE_GPP);
                    //音频编码格式
                    recorder.setAudioEncoder(MediaRecorder.
                    AudioEncoder.AMR_NB);
                    //音频文件保存的位置
                    recorder.setOutputFile(soundFile.getAbsolutePath());
                    //准备录音
                    recorder.prepare();
                    //开始录音
                    recorder.start();
                    //输出信息提示
                    System.out.println("开始录音");
                } catch (IOException e) {
                    // TODO Auto-generated catch block
                    e.printStackTrace();
                }
```

```java
            }
        });
        stop = (Button)findViewById(R.id.button2);
        stop.setOnClickListener(new OnClickListener() {
            public void onClick(View v) {
                // TODO Auto-generated method stub
                //音频文件存在且不为空
                if (soundFile != null && soundFile.exists()) {
                    //停止录音
                    recorder.stop();
                    //释放资源
                    recorder.release();
                    //记录器置为空
                    recorder = null;
                    //输出信息提示
                    System.out.println("停止录音");
                }
            }
        });
    }

    @Override
    protected void onDestroy() {
     // TODO Auto-generated method stub
     if (soundFile != null && soundFile.exists()) {
         //退出程序时停止录音
         recorder.stop();
         recorder.release();
         recorder = null;
     }
     super.onDestroy();
    }
}
```

AndroidManifest.xml：

```xml
<manifest xmlns:android="http://schemas.android.com/apk/res/android"
    package="com.example.recordsound"
    android:versionCode="1"
    android:versionName="1.0" >
    <uses-sdk
        android:minSdkVersion="8"
        android:targetSdkVersion="15" />
    <uses-permission android:name="android.permission.RECORD_AUDIO"/>
    <uses-permission android:name="android.permission.MOUNT_UNMOUNT_FILESYSTEMS"/>
    <uses-permission android:name="android.permission.WRITE_EXTERNAL_STORAGE"/>
    <application
        android:icon="@drawable/ic_launcher"
        android:label="@string/app_name"
        android:theme="@style/AppTheme" >
        <activity
            android:name=".RecordSoundActivity"
            android:label="@string/title_activity_record_sound" >
            <intent-filter>
                <action android:name="android.intent.action.MAIN" />
                <category android:name="android.intent.category.LAUNCHER" />
            </intent-filter>
        </activity>
```

```
        </application>
</manifest>
```

运行程序。单击"开始录音"按钮，LogCat 面板输出信息，提示开始录音；单击"停止录音"按钮，LogCat 面板输出信息，提示停止录音。如图 11.11 所示。

```
I  08-26 02:36:43.579   1241   1241   com.example.recordsound   System.out   开始录音
I  08-26 02:37:34.599   1241   1241   com.example.recordsound   System.out   停止录音
```

图 11.11　LogCat 面板输出的信息

打开 DDMS 视图的 File Explorer 面板，在 mnt/sdcard 路径下生成 sound.amr 音频文件，如图 11.12 所示。该文件由电脑的麦克风录制而成。

图 11.12　录制生成的音频文件

11.3.2　视频录制

MediaRecorder 除了可以录制音频以外，还可以录制视频。使用 MediaRecorder 录制视频与录制音频的步骤基本相同。只是录制视频不仅需要录制声音，还需要录制图像。录制图像用到的相关方法如表 11-4 所示。

表 11-4　录制图像的相关方法

方 法 名 称	方 法 说 明
public void setVideoEncoder(int video_encoder)	设置视频编码格式
public void setVideoFrameRate(int rate)	设置视频的帧率
public void setVideoSize(int width, int height)	设置视频的宽度和高度
public void setVideoSource(int video_source)	设置视频来源

【示例 11-8】　视频录制。新建项目 RecordVideo。在界面添加两个系统提供的 ImageButton、btn_star_big_on 按钮用来开始录制视频、btn_star_big_off 按钮用来停止录制视频。

逻辑代码如下：

```java
public class RecordVideoActivity extends Activity {
    //录像开关按钮
    private ImageButton on,off;
    //记录器
    private MediaRecorder recorder;
    //视频文件
    private File videoFile;
    //记录器是否正在录像
```

```java
private boolean isRecoding = false;
@Override
public void onCreate(Bundle savedInstanceState) {
    super.onCreate(savedInstanceState);
    setContentView(R.layout.activity_record_video);
    on = (ImageButton)findViewById(R.id.imageButton1);
    on.setOnClickListener(new OnClickListener() {
        public void onClick(View v) {
            // TODO Auto-generated method stub
            //如果无SDCard可用,则使用Toast提示
            if (!Environment.getExternalStorageState().equals
            (Environment.MEDIA_MOUNTED)) {
                Toast.makeText(RecordVideoActivity.this, "请插入SDCard!",
                        Toast.LENGTH_SHORT).show();
                return;
            }
            try {
                //创建保存录像的视频文件
                videoFile = new
                File(Environment.getExternalStorageDirectory().
                getCanonicalFile() + "/video.mp4");
                //创建记录器
                recorder = new MediaRecorder();
                recorder.setAudioSource(MediaRecorder.
                AudioSource.MIC);
                //视频来源于摄像头
                recorder.setVideoSource(MediaRecorder.VideoSource.
                CAMERA);
                //视频文件以MP4格式输出
                recorder.setOutputFormat(MediaRecorder.OutputFormat.
                MPEG_4);
                //音频以默认格式编码
                recorder.setAudioEncoder(MediaRecorder.AudioEncoder.
                DEFAULT);
                //图像以MP4格式编码
                recorder.setVideoEncoder(MediaRecorder.VideoEncoder.
                MPEG_4_SP);
                //图像宽度和高度
                recorder.setVideoSize(320, 240);
                //每秒4帧
                recorder.setVideoFrameRate(4);
                //获取视频文件绝对路径
                recorder.setOutputFile(videoFile.getAbsolutePath());
                //准备录像
                recorder.prepare();
                //开始录像
                recorder.start();
                //正在录像
                isRecoding = true;
                //输出信息提示
                System.out.println("开始录像");
            } catch (IOException e) {
                // TODO Auto-generated catch block
                e.printStackTrace();
            }
        }
    });
    off = (ImageButton)findViewById(R.id.imageButton2);
    off.setOnClickListener(new OnClickListener() {
```

```java
        public void onClick(View v) {
            // TODO Auto-generated method stub
            //如果正在录像
            if (isRecoding) {
                //停止录像
                recorder.stop();
                //释放资源
                recorder.release();
                //记录器置为空
                recorder = null;
                //输出提示信息
                System.out.println("停止录像");
            }
        }
    });
}
}
```

AndroidManifest.xml：

```xml
<manifest xmlns:android="http://schemas.android.com/apk/res/android"
    package="com.example.recordsound"
    android:versionCode="1"
    android:versionName="1.0" >
    <uses-sdk
        android:minSdkVersion="8"
        android:targetSdkVersion="15" />

    <uses-permission android:name="android.permission.CAMERA"/>
    <uses-permission android:name="android.permission.RECORD_AUDIO"/>
    <uses-permission android:name="android.permission.MOUNT_UNMOUNT_FILESYSTEMS"/>
    <uses-permission android:name="android.permission.WRITE_EXTERNAL_STORAGE"/>

    <application
        android:icon="@drawable/ic_launcher"
        android:label="@string/app_name"
        android:theme="@style/AppTheme" >
        <activity
            android:name=".RecordSoundActivity"
            android:label="@string/title_activity_record_sound" >
            <intent-filter>
                <action android:name="android.intent.action.MAIN" />
                <category android:name="android.intent.category.LAUNCHER" />
            </intent-filter>
        </activity>
    </application>
</manifest>
```

运行程序。单击 btn_star_big_on 按钮开始录像，单击 btn_star_big_off 停止录像。LogCat 面板输出的提示信息如图 11.13 所示。

```
I    08-27 01:36:18.654    1686    1686    com.example.recordvideo    System.out    开始录像
I    08-27 01:37:06.854    1686    1686    com.example.recordvideo    System.out    停止录像
```

图 11.13 LogCat 面板输出的提示信息

打开 DDMS 视图的 File Explorer 面板，在 mnt/sdcard 路径下生成 video.mp4 视频文件，

如图 11.14 所示。

图 11.14　录制生成的视频文件

注意：由于模拟器没有摄像头硬件支持，所以无法录制图像。建议采用有摄像头硬件支持的真机测试该程序。

11.4　相机 Camera

在 Android 多媒体应用开发中，我们可以调用系统功能拍照，同样需要硬件设备（摄像头）的支持。Android 支持的图像格式有 JPEG、GIF、PNG 和 BMP。

【示例 11-9】　调用系统功能照相。新建项目 Camera，在界面添加一个"拍照"按钮，单击后启动系统拍照功能；添加一个 ImageView，显示拍摄的照片。

逻辑代码如下：

```java
public class CameraActivity extends Activity{
    //显示照片
    private ImageView imageView;
    //拍照按钮
    private Button takeButton;
    //照片对象
    private Bitmap myBitmap;

    /** Called when the activity is first created. */
    @Override
    public void onCreate ( Bundle savedInstanceState )
    {
        super.onCreate(savedInstanceState);
        setContentView(R.layout.activity_camera);
        takeButton = (Button)findViewById(R.id.button1);
        takeButton.setOnClickListener(new OnClickListener() {
            public void onClick(View v) {
                // TODO Auto-generated method stub
                //创建系统拍照 Intent
                Intent getImageByCamera = new Intent("android.media.action.IMAGE_CAPTURE");
                //调用 startActivityForResult(Intent intent, int requestCode)方法启动系统拍照界面
                startActivityForResult(getImageByCamera, 1);
            }
        });
```

```java
    }
    @Override
    //调用 onActivityResult(int requestCode, int resultCode, intent data)
       方法,通过判断结果码获得返回的照片
    protected void onActivityResult ( int requestCode , int resultCode ,
    Intent data ){
        // TODO Auto-generated method stub
        super.onActivityResult(requestCode, resultCode, data);
        //结果码匹配成功
        if (requestCode == 1){
            try{
                super.onActivityResult(requestCode, resultCode, data);
                //获取拍摄的图片
                Bundle extras = data.getExtras();
                //将图片转换为 Bitmap 格式
                myBitmap = (Bitmap) extras.get("data");
            } catch ( Exception e ){
            // TODO Auto-generated catch block
            e.printStackTrace();
            }
            imageView = (ImageView) findViewById(R.id.imageView1);
            //将图片显示在 ImageView 中
            imageView.setImageBitmap(myBitmap);
        }
    }
}
```

AndroidManifest.xml：

```xml
<manifest xmlns:android="http://schemas.android.com/apk/res/android"
    package="com.example.camera"
    android:versionCode="1"
    android:versionName="1.0" >
    <uses-sdk
        android:minSdkVersion="8"
        android:targetSdkVersion="15" />

    <uses-permission android:name="android.permission.CAMERA"/>
    <uses-feature android:name = "android.hardware.camera" />
    <uses-feature android:name = "android.hardware.camera.autofocus" />

    <application
        android:icon="@drawable/ic_launcher"
        android:label="@string/app_name"
        android:theme="@style/AppTheme" >
        <activity
            android:name=".CameraActivity"
            android:label="@string/title_activity_camera" >
            <intent-filter>
                <action android:name="android.intent.action.MAIN" />
                <category android:name="android.intent.category.LAUNCHER" />
            </intent-filter>
        </activity>
    </application>
</manifest>
```

运行程序，效果如图 11.15 所示。

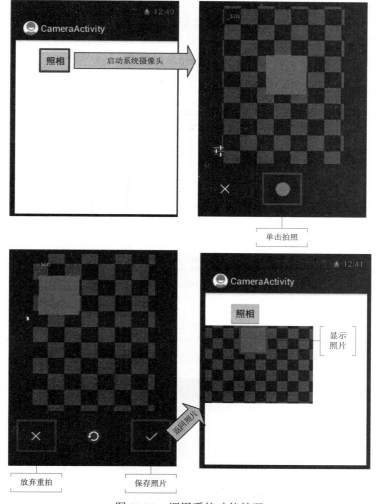

图 11.15 调用系统功能拍照

提示：由于程序在模拟器中运行，所以获取的图片如图 11.15 所示。如果在真机中测试，则可以启动设备摄像头，进行拍照。

11.5 小　　结

本章主要介绍了 Android 系统中多媒体的应用开发。其中，使用 MediaPlayer 播放音频、使用 VideoView 播放视频的开发都比较简单，较容易掌握。音频、视频，以及图像的采集是本章的难点，需要读者多多练习，以便熟练掌握。

11.6 习　　题

1. 新建项目 Music。导入多个 MP3 音频文件到 SDCard 中，然后使用 Mediaplayer 类的相关方法，播放指定的音乐。

【分析】本题目主要考查读者对使用 MediaPlayer 播放存储在文件系统中的标准音频文件的掌握。注意播放的标准音频文件保存在 SDCard 中。调用 MediaPlayer.setDataSource (String path)播放指定路径下的音频文件。可以参考 11.1.3 节的内容。

【核心代码】本题的核心代码如下所示。

```java
try {
    mPlayer.reset();
    mPlayer.setDataSource("/mnt/sdcard/miracle.mp3");
    mPlayer.prepare();
    mPlayer.start();
} catch (IllegalArgumentException e) {
    // TODO Auto-generated catch block
    e.printStackTrace();
} catch (SecurityException e) {
    // TODO Auto-generated catch block
    e.printStackTrace();
} catch (IllegalStateException e) {
    // TODO Auto-generated catch block
    e.printStackTrace();
} catch (IOException e) {
    // TODO Auto-generated catch block
    e.printStackTrace();
}
```

2. 新建项目 VideoView，在界面添加 VideoView 控件，使用 VideoView 播放存储在 SDcard 中的视频文件。使用 MediaController 媒体控制器，控制视频的播放，如图 11.16 所示。

图 11.16　播放视频文件

【分析】本题目主要考查读者对使用 VideoView 播放视频文件的掌握。需要注意的是，播放的视频文件需要事先导入到 SDcard 目录下。可以参考 11.2.2 节的内容。

【核心代码】本题的核心代码如下所示。

```java
setContentView(R.layout.activity_video_view);
videoView = (VideoView)findViewById(R.id.videoView1);
MediaController mController = new MediaController(this);
videoView.setVideoPath("/mnt/sdcard/video.3gp");
videoView.setMediaController(mController);
mController.setMediaPlayer(videoView);
```

3. 新建项目 Camera，调用系统功能拍照，获取拍摄的照片显示在 ImageView 控件中。

【分析】本题目主要考查读者对相机 Camera 调用的掌握。注意调用的时候，需要建立

Intent。可以参考 11.4 节的内容。

【核心代码】本题的核心代码如下所示。

```
Intent getImageByCamera = new Intent ("android.media.action.IMAGE_CAPTURE");
    startActivityForResult(getImageByCamera, 1);
@ Override
    protected void onActivityResult ( int requestCode , int resultCode ,
    Intent data ){
        // TODO Auto-generated method stub
        super.onActivityResult(requestCode, resultCode, data);
            if (requestCode == 1){
                try{
                    super.onActivityResult(requestCode, resultCode, data);
                    Bundle extras = data.getExtras();
                    myBitmap = (Bitmap) extras.get("data");
                } catch ( Exception e ){
                    // TODO Auto-generated catch block
                    e.printStackTrace();
                }
            imageView = (ImageView) findViewById(R.id.imageView1);
            imageView.setImageBitmap(myBitmap);
        }
    }
```

第 12 章　Android 感应检测——Sensor

在 Android 系统中，提供了对传感器的支持。传感器在 Android 的应用中起到了非常重要的作用，可以实现一些我们意想不到的功能，比如音乐键盘、火灾报警、地震仪等。本章将介绍一些传感器的开发及应用。

12.1　Sensor 简介

在开发传感器应用之前，首先了解传感器的开发过程。要测试感应检测 Sensor 的功能，必须在装有 Android 系统的真机设备上进行。

12.1.1　Sensor 种类

Android 中支持的 Sensor 种类如表 12-1 所示。

表 12-1　系统传感器

感 应 检 测	说　　明
TYPE_ACCELEROMETER	加速度传感器
TYPE_AMBIENT_TEMPERATURE	温度传感器
TYPE_GRAVITY	重力传感器
TYPE_GYROSCOPE	回转仪传感器
TYPE_LIGHT	光传感器
TYPE_LINEAR_ACCELERATION	线性加速度传感器
TYPE_MAGNETIC_FIELD	磁场传感器
TYPE_PRESSURE	压力传感器
TYPE_PROXIMITY	接近传感器
TYPE_RELATIVE_HUMIDITY	相对湿度传感器
TYPE_ROTATION_VECTOR	旋转矢量传感器

12.1.2　Sensor 开发

传感器应用程序的开发分为以下几个步骤。

（1）调用 Context.getSystemService(SENSOR_SERVICE)方法获取传感器管理服务。实例化方法如下：

```
SensorManager manager = (SensorManager)getSystemService(SENSOR_SERVICE);
```

（2）调用 SensorManager 的 getDefaultSensor(int type)方法，获取指定类型的传感器。方法格式如下：

```
SensorManager.getDefaultSensor ( int type )
```

（3）在 Activity 的 onResume()中，调用 SensorManager 的 registerListener(SensorEventListener listener, Sensor sensor, int rate)方法注册监听。

```
SensorManager.registerListener(           // 注册监听器
    SensorEventListener listener,         // 监听传感器事件
    Sensor sensor,                        // 传感器对象
    int rate)                             // 延迟时间精密度
```

rate 支持的参数介绍如下。

- Sensor.manager.SENSOR_DELAY_FASTEST：延迟 0ms；
- Sensor.manager.SENSOR_DELAY_GAME：延迟 20ms，适合游戏的频率；
- Sensor.manager.SENSOR_DELAY_UI：延迟 60ms，适合普通界面的频率；
- Sensor.manager.SENSOR_DELAY_NORMAL：延迟 200ms，正常频率。

（4）实现 SensorEventListener 接口中下列两个方法，监听并取得传感器 Sensor 的状态。

```
public abstract void onAccuracyChanged(Sensor sensor, int accuracy)
                                                              // 监听传感器精度变化
public abstract void onSensorChanged(SensorEvent event)       // 监听传感器值变化
```

12.1.3　Sensor 真机测试

由于我们以往使用的模拟器不支持传感器感应功能，所以本章示例都在真机中进行测试（笔者使用的是支持 Android 4.0 的手机）。下面介绍真机测试步骤。

（1）设置手机为 USB 调试模式。依次选择"设置"|"应用程序"|"开发"|"USB 调试"命令，如图 12.1 所示。

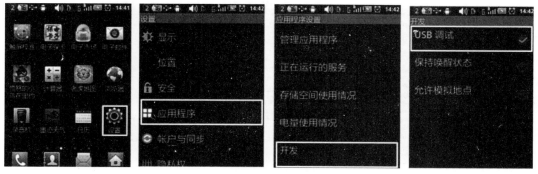

图 12.1　设置 USB 调试

注意：使用的设备不同，USB 调试模式的设置也不同。请读者参考该步骤，自行设置。

（2）用 USB 数据线连接手机到计算机。使用 360 手机助手下载驱动，确保连接成功。

（3）在 DOS 窗口下执行 adb devices 命令，查看手机是否已经连接成功。如果输出手机信息，则表示连接成功。如图 12.2 所示。

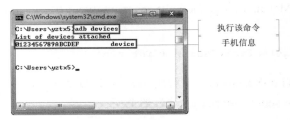

图 12.2　手机连接成功

（4）执行真机调试操作。在 Eclipse 中，右击要运行的项目文件，依次选择 Run As|Run Configurations…命令，打开 Run Configurations 对话框。在对话框的左侧选中项目名称。在右侧打开 Target 面板，选择 Launch on all compatible devices/AVD's 单选按钮，再通过下拉菜单选择 Active devices 选项。然后，单击 Apply 按钮应用。最后，单击 Run 按钮，程序在真机中运行。步骤如图 12.3 所示。

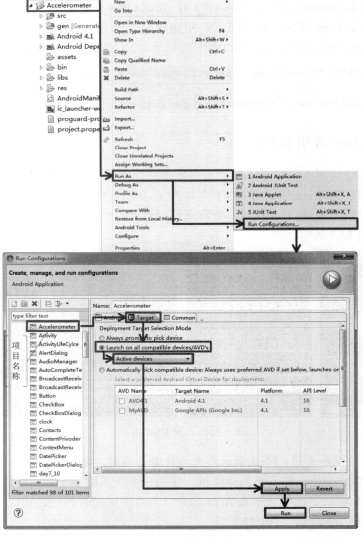

图 12.3　真机测试

12.1.4 Sensor 信息检测

感应检测 Sensor 的硬件组件由不同的厂商提供。不同的 Sensor 设备组件，所检测的事件也不同。可以使用 Sensor 类的 getXXX()方法，检测设备所支持的 Sensor 的相关信息，如表 12-2 所示。

表 12-2 Sensor 的相关方法

方 法 名 称	方 法 说 明
public float getMaximumRange()	获取 Sensor 最大值
public int getMinDelay()	获取 Sensor 的最小延迟
public String getName()	获取 Sensor 名称
public float getPower()	获取 Sensor 使用时所耗功率
public float getResolution()	获取 Sensor 的精度
public int getType()	获取 Sensor 类型
public String getVendor()	获取 Sensor 供应商信息
public int getVersion()	获取 Sensor 版本号信息

12.2 常用系统传感器

Android 系统的亮点之一就是对传感器的应用，Android 系统提供了 10 余种传感器，本节将选择几种常用的传感器来介绍其开发及应用过程。

12.2.1 方向传感器

方向传感器（Orientation）简称为 O-sensor，主要感应方位的变化。现在已经被 SensorManager.getOrientation()所取代，我们可以通过磁力计 MagneticField 和加速度传感器 Accelerometer 来获得方位信息。该传感器同样捕获三个参数，分别代表手机沿传感器坐标系的 X 轴、Y 轴和 Z 轴转过的角度。

- values[0]：azimuth 方向角，但用（磁场+加速度）得到的数据范围是-180～180。也就是说，0 表示正北，90 表示正东，180/-180 表示正南，-90 表示正西。而直接通过方向感应器数据范围是 0～359。其中，0 表示正北，90 表示正东，180 表示正南，270 表示正西。
- values[1]：pitch 倾斜角，围绕 X 轴的旋转角。由静止状态开始，前后翻转，取值范围为-180 度～180 度。
- values[2]：roll 旋转角，围绕 Y 轴的旋转角。由静止状态开始，左右翻转，取值范围为-90 度～90 度。

【示例 12-1】方向传感器的开发。新建项目 Orientation。在布局界面添加 3 个文本框：xView 用于显示倾斜角，yView 用于显示旋转角，zView 用于显示方向角。用真机测试程序，查看运行效果。

```
//实现 SensorEventListener 接口,并覆盖该接口中的 onAccuracyChanged 方法和
  onSensorChanged 方法
```

```java
public class OrientationActivity extends Activity implements SensorEventListener{
    //分别显示 X、Y、Z 轴方向上的方向角度
    private TextView xView,yView,zView;
    // SensorManager 对象
    private SensorManager manager;
    //加速度传感器对象 aSensor,磁场传感器对象 mSensor
    private Sensor aSensor,mSensor;
    //加速度数组
    float[] accelerometerValues=new float[3];
    //磁场数组
    float[] magneticFieldValues=new float[3];
    //转换矩阵,保存方位信息
    float[] values=new float[3];
    //旋转矩阵,保存加速度和磁场数据
    float[] rotate=new float[9];
    @Override
    public void onCreate(Bundle savedInstanceState) {
        super.onCreate(savedInstanceState);
        setContentView(R.layout.activity_orientation);

        xView = (TextView)findViewById(R.id.textView1);
        yView = (TextView)findViewById(R.id.textView2);
        zView = (TextView)findViewById(R.id.textView3);
        //获取传感器管理服务
        manager = (SensorManager)getSystemService(SENSOR_SERVICE);
        //获取加速度传感器
        aSensor = manager.getDefaultSensor(Sensor.TYPE_ACCELEROMETER);
        //获取磁场传感器
        mSensor = manager.getDefaultSensor(Sensor.TYPE_MAGNETIC_FIELD);
    }
    @Override
    protected void onResume() {
        // TODO Auto-generated method stub
        //为加速度传感器注册监听
        manager.registerListener(
                this,
                aSensor,
                manager.SENSOR_DELAY_UI);
        //为磁场传感器注册监听
        manager.registerListener(
                this,
                mSensor,
                manager.SENSOR_DELAY_UI);
        super.onResume();
    }
    public void onAccuracyChanged(Sensor sensor, int accuracy) {
        // TODO Auto-generated method stub

    }
    public void onSensorChanged(SensorEvent event) {
        // TODO Auto-generated method stub
        //获取加速度传感器值
        if(event.sensor.getType()==Sensor.TYPE_ACCELEROMETER){
            accelerometerValues=event.values;
        }

        //获取磁场传感器值
        if(event.sensor.getType()==Sensor.TYPE_MAGNETIC_FIELD){
            magneticFieldValues=event.values;
```

```
        }
        //填充旋转矩阵
        SensorManager.getRotationMatrix(
        //旋转矩阵
        rotate,
        null,
        //从加速度感应器获取的数据
        accelerometerValues,
        //从磁场感应器获取的数据
        magneticFieldValues);
        //将弧度转换为角度
        SensorManager.getOrientation(rotate, values);
        //获取各角度显示
        values[0]=(float)Math.toDegrees(values[0]);
        zView.setText("z="+values[0]);
        values[1]=(float)Math.toDegrees(values[1]);
        xView.setText("x="+values[1]);
        values[2]=(float)Math.toDegrees(values[2]);
        yView.setText("y="+values[2]);
    }
}
```

12.2.2 磁场传感器

磁场传感器（MagneticField）简称为 M-sensor，该传感器主要读取的是磁场的变化，通过该传感器便可开发出指南针、罗盘等磁场应用。磁场传感器读取的数据同样是空间坐标系三个方向的磁场值，其数据单位为 uT，即微特斯拉。下面通过对 12.2.1 节方向传感器案例的更改，来完成磁场数据的读取。

【示例 12-2】 磁场传感器的开发。新建项目 MagneticField。在布局界面添加 3 个文本框：xView 用于显示 X 轴方向上的磁场分量，yView 用于显示 Y 轴方向上的磁场分量，zView 用于显示 Z 轴上的磁场分量。用真机测试程序，查看运行效果。
逻辑代码如下：

```
//实现 SensorEventListener 接口,并覆盖该接口中的 onAccuracyChanged 方法和
  onSensorChanged 方法
public class MagneticFieldActivity extends Activity implements
SensorEventListener{
    //分别显示 X、Y、Z 轴方向上的磁场分量
    private TextView xView,yView,zView;
    // SensorManager 对象
    private SensorManager manager;
    //传感器对象
    private Sensor sensor;
    @Override
    public void onCreate(Bundle savedInstanceState) {
        super.onCreate(savedInstanceState);
        setContentView(R.layout.activity_magnetic_field);

        xView = (TextView)findViewById(R.id.textView1);
        yView = (TextView)findViewById(R.id.textView2);
        zView = (TextView)findViewById(R.id.textView3);
        //获取传感器管理服务
        manager = (SensorManager)getSystemService(SENSOR_SERVICE);
```

```java
        //获取磁场传感器
        sensor = manager.getDefaultSensor(Sensor.TYPE_MAGNETIC_FIELD);
    }
@Override
//覆盖onResume()方法
protected void onResume() {
    // TODO Auto-generated method stub
    //注册监听
    manager.registerListener(
        //监听器对象
        this,
        //磁场传感器对象
        sensor,
        //延迟60ms
        manager.SENSOR_DELAY_UI);
    super.onResume();
}
public void onAccuracyChanged(Sensor sensor, int accuracy) {
    // TODO Auto-generated method stub
}
public void onSensorChanged(SensorEvent event) {
    // TODO Auto-generated method stub
//显示磁场分量值
    float [] values = event.values;
    xView.setText("X方向上的磁场分量为: " + values[0]);
    yView.setText("Y方向上的磁场分量为: " + values[1]);
    zView.setText("Z方向上的磁场分量为: " + values[2]);
    }
}
```

12.2.3 重力传感器

重力传感器（Gravity）简称 GV-sensor，主要用于输出重力数据。在地球上，重力数值为 9.8，单位是 m/s^2。坐标系统与加速度传感器坐标系相同。当设备复位时，重力传感器的输出与加速度传感器相同。

【示例 12-3】重力传感器的开发。新建项目 Gravity。在布局界面添加 3 个文本框：xView 用于显示 X 轴方向上的重力，yView 用于显示 Y 轴方向上的重力，zView 用于显示 Z 轴上的重力。用真机测试程序，查看运行效果。

```java
//实现SensorEventListener接口,并覆盖该接口中的onAccuracyChanged方法和
  onSensorChanged方法
public class GravityActivity extends Activity implements SensorEventListener{
    //分别显示X、Y、Z轴方向上的重力值
    private TextView xView,yView,zView;
    //SensorManager对象
    private SensorManager manager;
    //传感器对象
    private Sensor sensor;
    @Override
    public void onCreate(Bundle savedInstanceState) {
        super.onCreate(savedInstanceState);
        setContentView(R.layout.activity_gravtiy);
```

```
        xView = (TextView)findViewById(R.id.textView1);
        yView = (TextView)findViewById(R.id.textView2);
        zView = (TextView)findViewById(R.id.textView3);
        //获取传感器管理服务
        manager = (SensorManager)getSystemService(SENSOR_SERVICE);
        //获取重力传感器
        sensor = manager.getDefaultSensor(Sensor.TYPE_GRAVITY);
    }

@Override
//覆盖 onResume()方法
protected void onResume() {
    // TODO Auto-generated method stub
    //注册监听
    manager.registerListener(
    //监听器对象
    this,
    //重力传感器对象
            sensor,
            //延迟 60ms
            manager.SENSOR_DELAY_UI);
    super.onResume();
}

    public void onAccuracyChanged(Sensor sensor, int accuracy) {
        // TODO Auto-generated method stub
        
    }
    public void onSensorChanged(SensorEvent event) {
        // TODO Auto-generated method stub
        //显示重力值
        float [] values = event.values;
        xView.setText("X 方向上的重力值为: " + values[0]);
        yView.setText("Y 方向上的重力值为: " + values[1]);
        zView.setText("Z 方向上的重力值为: " + values[2]);
    }
}
```

12.2.4 加速度传感器

加速度传感器（Accelerometer）简称 G-sensor，主要用于感应设备的运动。该传感器捕获三个参数，分别表示空间坐标系中 X、Y、Z 轴方向上的加速度减去重力加速度在相应轴上的分量，其单位均为 m/s²。

如图 12.4 所示，传感器的坐标系与手机屏幕中的坐标系不同。传感器坐标系以屏幕左下角为原点，X 轴沿着屏幕向右，Y 轴沿屏幕向上，Z 轴垂直于手机屏幕向上。

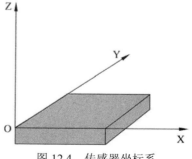

图 12.4　传感器坐标系

【示例 12-4】 加速度传感器的开发。新建项目 Accelerometer，在布局界面添加 3 个文本框：xView 用于显示 X 轴方向上的加速度，yView 用于显示 Y 轴方向上的加速度，zView 用于显示 Z 轴上的加速度。

逻辑代码如下:

```java
//实现 SensorEventListener 接口,并覆盖该接口中的 onAccuracyChanged 方法和
  onSensorChanged 方法
public class AccelerometerActivity extends Activity implements
SensorEventListener{
    //分别显示 X、Y、Z 轴方向上的加速度
    private TextView xView,yView,zView;
    // SensorManager 对象
    private SensorManager manager;
    //传感器对象
    private Sensor sensor;
    @Override
    public void onCreate(Bundle savedInstanceState) {
        super.onCreate(savedInstanceState);
        setContentView(R.layout.activity_accelerometer);

        xView = (TextView)findViewById(R.id.textView1);
        yView = (TextView)findViewById(R.id.textView2);
        zView = (TextView)findViewById(R.id.textView3);
        //获取传感器管理服务
        manager = (SensorManager)getSystemService(SENSOR_SERVICE);
        //获取加速度传感器
        sensor = manager.getDefaultSensor(Sensor.TYPE_ACCELEROMETER);
    }
@Override
//覆盖 onResume()方法
protected void onResume() {
    // TODO Auto-generated method stub
    //注册监听
    manager.registerListener(
        //监听器对象
        this,
            //加速度传感器对象
            sensor,
            //延迟 60ms
            manager.SENSOR_DELAY_UI);
    super.onResume();
    }
    public void onAccuracyChanged(Sensor sensor, int accuracy) {
        // TODO Auto-generated method stub
    }
    public void onSensorChanged(SensorEvent event) {
        // TODO Auto-generated method stub
        //显示加速度值
        float [] values = event.values;
        xView.setText("X 方向上的加速度分量为: " + values[0]);
        yView.setText("Y 方向上的加速度分量为: " + values[1]);
        zView.setText("Z 方向上的加速度分量为: " + values[2]);
    }
}
```

在真机运行程序,运行操作参考 12.1.3 节的内容。运行效果如图 12.5 所示。

12.2.5 光传感器

光传感器(Light),主要用来检测设备周围光线强度。光强单位是勒克斯(lux),其物理意义是照射到单位面积上的光通量。光传感器的开发与之前介绍过的各种传感器的开发步骤基本相同,只是监测的是 SENSOR_LIGHT,即捕捉光的强度。

第 12 章　Android 感应检测——Sensor

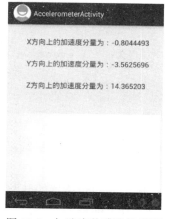

图 12.5　加速度传感器效果图

【示例 12-5】　光传感器的开发。新建项目 Light。在界面添加一个文本框，用于显示光线强度。用真机测试程序，查看运行效果。

逻辑代码如下：

```java
//实现 SensorEventListener 接口,并覆盖该接口中的 onAccuracyChanged 方法和
  onSensorChanged 方法
public class LightActivity extends Activity implements SensorEventListener{
    //显示光线强度
    private TextView view;
    // SensorManager 对象
    private SensorManager manager;
    //传感器对象
    private Sensor sensor;
    @Override
    public void onCreate(Bundle savedInstanceState) {
        super.onCreate(savedInstanceState);
        setContentView(R.layout.activity_light);
        view = (TextView)findViewById(R.id.textView1);
        //获取传感器管理服务
        manager = (SensorManager)getSystemService(SENSOR_SERVICE);
        //获取亮度传感器
        sensor = manager.getDefaultSensor(Sensor.TYPE_LIGHT);
    }
@Override
//覆盖 onResume()方法
protected void onResume() {
    // TODO Auto-generated method stub
    //注册监听
    manager.registerListener(
    //监听器对象
    this,
    //亮度传感器对象
    sensor,
    //延迟 60ms
    manager.SENSOR_DELAY_UI);
    super.onResume();
    }
    public void onAccuracyChanged(Sensor sensor, int accuracy) {
        // TODO Auto-generated method stub
```

· 257 ·

```
    }
    public void onSensorChanged(SensorEvent event) {
        // TODO Auto-generated method stub
//显示光线强度
        view.setText("光的强度为: " + event.values[0]);
    }
}
```

12.3 小　　结

本章主要讲解了 Android 系统中，常用传感器的开发及应用。本章重点在于如何在真机上测试自己开发的应用程序，我们开发的程序最终都是运行在真机上，掌握真机测试十分重要。读者掌握这些传感器的使用之后，可以自主开发一些传感器应用程序，更加深入地学习传感器的应用。

12.4 习　　题

新建项目 Sensor，通过程序测验自己使用的设备支持哪几种传感器。

【分析】本题目主要考查读者对传感器的掌握。通过该题，读者可以了解到自己的设备所支持的传感器，以便之后有目的的练习各传感器的使用。

【核心代码】本题的核心代码如下所示。

```
manager = (SensorManager)getSystemService(SENSOR_SERVICE);
List<Sensor> allSensors = manager.getSensorList(Sensor.TYPE_ALL);
view.setText("经检测该设备有" + allSensors.size() + "个传感器,他们分别是: \n");
for (Sensor s : allSensors) {
    String tempString = "\n" +
            " 设备名称: " + s.getName() + "\n" +
            " 设备版本: " + s.getVersion() + "\n" +
            " 供应商: " + s.getVendor() + "\n";
    switch (s.getType()) {
    case Sensor.TYPE_ACCELEROMETER:
        view.setText(view.getText().toString() + s.getType() + " 加速度传感器" + tempString);
    case Sensor.TYPE_MAGNETIC_FIELD:
        view.setText(view.getText().toString() + s.getType() + " 磁场传感器" + tempString);
    case Sensor.TYPE_PRESSURE:
        view.setText(view.getText().toString() + s.getType() + " 压力传感器" + tempString);
    case Sensor.TYPE_GRAVITY:
        view.setText(view.getText().toString() + s.getType() + " 重力传感器" + tempString);
    case Sensor.TYPE_LIGHT:
        view.setText(view.getText().toString() + s.getType() + " 光传感器" + tempString);
    }
}
```

第 13 章　手势识别和无线网络

手势识别（Android Gesture）是用来侦测、处理手势相关动作的技术。Gesture 大致可以分为两类：一类是触摸屏手势，另一类是输入法手势。无线通信（Wireless Communication）是利用电磁波信号在自由空间中传播的特性进行信息交换的一种通信方式。Android 中最常用的无线通信就是 Wi-Fi 和蓝牙（Bluetooth）。下面将分别来介绍这两种应用。

13.1　触摸屏手势

触摸屏手势比较简单，通常是按下、抬起、滑动和翻页这 4 种。Android 系统为我们提供了手势识别工具 GestureDetector，当我们接收到用户触摸消息时，将这个消息交给 GestureDetector 去加工，我们通过设置监听器获得 GestureDetector 处理后的手势。

13.1.1　GestureDetector 简介

GestureDetector 提供了两个监听器接口，OnGestureListener 处理单击类消息，如表 13-1 所示；OnDoubleTapListener 处理双击类消息，如表 13-2 所示。

表 13-1　OnGestureListener 的接口

接 口 名 称	接 口 说 明
onDown(MotionEvent e)	单击，触摸屏按下时立刻触发
onSingleTapUp(MotionEvent e)	抬起，手指离开触摸屏时触发（长按、滚动、滑动时，不会触发这个手势）
onShowPress(MotionEvent e)	长按，触摸屏按下后既不抬起也不移动，过一段时间后触发
onShowPress(MotionEvent e)	短按，触摸屏按下后片刻后抬起，会触发这个手势，如果迅速抬起则不会
onScroll(MotionEvent e1, MotionEvent e2, float distanceX, float distanceY)	滚动，触摸屏按下后移动
onFling(MotionEvent e1, MotionEvent e2, float velocityX, float velocityY)	滑动，触摸屏按下后快速移动并抬起，会先触发滚动手势，跟着触发一个滑动手势

表 13-2　OnDoubleTapListener 的接口

接 口 名 称	接 口 说 明
onDoubleTap(MotionEvent e)	双击，手指在触摸屏上迅速点击第二下时触发
onDoubleTapEvent(MotionEvent e)	双击的按下跟抬起各触发一次
onSingleTapConfirmed(MotionEvent e)	单击确认，即很快地按下并抬起，但并不连续点击第二下

下面是各种操作返回的手势序列,数值 0 表示触摸屏按下,1 表示抬起。

- 单击:down 0、single up 1、single conf 0;
- 短按:down 0、show 0、single up 1;
- 长按:down 0、show 0、ong 0;
- 双击:down 0、single up 1、double 0、double event 0、down 0、double event 1;
- 滚动:down 0、show 0、scrool 2;
- 滑动:down 0、show 0、scrool 2、fling1。

13.1.2 触摸屏手势应用

有时候我们并不需要处理上面所有的手势。为方便起见,Android 提供了另外一个类 SimpleOnGestureListener 实现了如上接口,只需要继承 SimpleOnGestureListener,然后重载感兴趣的手势即可。

【示例 13-1】 触摸屏手势开发。新建项目 Gesture。再新建类 MyGestureListener 继承于 SimpleOnGestureListener,并覆盖 SimpleOnGestureListener 类中的方法,当用户以不同的方式触摸屏幕,Toast 弹出对应的消息提示。

MyGestureListener:

```java
public class MyGestureListener extends SimpleOnGestureListener {
    // MyGestureListener 构造方法参数
    private Context mContext;
    // MyGestureListener 构造方法
    MyGestureListener(Context context) {
        mContext = context;
    }
@Override
//单击
    public boolean onDown(MotionEvent e) {
        Toast.makeText(mContext, "DOWN " + e.getAction(), Toast.LENGTH_
        SHORT).show();
        return false;
    }
@Override
//短按
    public void onShowPress(MotionEvent e) {
        Toast.makeText(mContext, "SHOW " + e.getAction(), Toast.LENGTH_
        SHORT).show();
    }
@Override
//抬起
    public boolean onSingleTapUp(MotionEvent e) {
        Toast.makeText(mContext, "SINGLE UP " + e.getAction(), Toast.LENGTH_
        SHORT).show();
        return false;
    }
@Override
//滚动
    public boolean onScroll(MotionEvent e1, MotionEvent e2,
        float distanceX, float distanceY) {
        Toast.makeText(mContext, "SCROLL " + e2.getAction(), Toast.LENGTH_
        SHORT).show();
        return false;
    }
```

```java
@Override
//长按
    public void onLongPress(MotionEvent e) {
        Toast.makeText(mContext, "LONG " + e.getAction(), Toast.LENGTH_
        SHORT).show();
    }
@Override
//滑动
    public boolean onFling(MotionEvent e1, MotionEvent e2, float velocityX,
        float velocityY) {
        Toast.makeText(mContext, "FLING " + e2.getAction(), Toast.LENGTH_
        SHORT).show();
        return false;
    }
@Override
//双击
    public boolean onDoubleTap(MotionEvent e) {
        Toast.makeText(mContext, "DOUBLE " + e.getAction(), Toast.LENGTH_
        SHORT).show();
        return false;
    }
@Override
//双击的按下跟抬起各触发一次
    public boolean onDoubleTapEvent(MotionEvent e) {
        Toast.makeText(mContext, "DOUBLE EVENT " + e.getAction(),
        Toast.LENGTH_SHORT).show();
        return false;
    }
@Override
//单击确认
    public boolean onSingleTapConfirmed(MotionEvent e) {
        Toast.makeText(mContext, "SINGLE CONF " + e.getAction(), Toast.
        LENGTH_SHORT).show();
        return false;
    }
}
```

Gesture2Activity：

```java
public class Gesture2Activity extends Activity {
    //声明 GestureDetector 对象
    private GestureDetector mGestureDetector;
    @Override
    public void onCreate(Bundle savedInstanceState) {
        super.onCreate(savedInstanceState);
        setContentView(R.layout.activity_gesture2);
        //实例化 GestureDetector 对象
        mGestureDetector = new GestureDetector(
        //上下文环境
        this,
        //绑定触摸屏监听器
        new MyGestureListener(this));
    }
@Override
//覆盖 onTouchEvent()方法
public boolean onTouchEvent(MotionEvent event) {
        //返回触摸屏事件
        return mGestureDetector.onTouchEvent(event);
    }
```

}

运行程序，查看运行效果，如图 13.1 所示。

图 13.1　触摸屏手势效果图

13.2　输入法手势

输入法手势就是在触摸屏上手绘一个形状，这个形状可以由一个或者多个笔画构成。创建完成之后，系统会自动保存手势动作在 mnt/sdcard/gestures 里面。把 gestures 文件复制到工程/res/raw 下，就可以在项目里面使用这些手势了。

13.2.1　Gesture 相关类简介

开发 Gesture 应用程序，可能会用到以下类和接口，如表 13-3 所示。

表 13-3　Gesture 相关类与接口

名　　称	说　　明
Gesture	代表一个手势对象
GestureLibrary	Gesture 库
GestureLibraries	GestureLibrary 的 Factory 库

续表

名 称	说 明
GestureOverlayView	Gesture 输入的透明性重叠层，可以放在其他 Widget 上面
OnGesturePerformedListener	设置在 GestureOverlayView 上的关于 Gesture 的监听
Prediction	Gesture 的预报，有 name 和 score 两个属性。name 表示手势名称，score 表示相似度，数值越大越相似，一般认为大于 1 即可

13.2.2 输入法手势应用

本节来开发输入法手势程序。首先打开模拟器中的 Gestures Builder 程序，单击 Add gesture 按钮，创建几个手势，并为手势命名。创建成功后，程序弹出 Toast 消息 "Gestures saved in /mnt/sdcard/gestures"，提示手势文件保存在/mnt/sdcard 路径下。创建过程如图 13.2 所示。

图 13.2　创建手势

【示例 13-2】 输入法手势程序开发。新建项目 Gesture，将 gestures 文件导入到项目的 res/raw 文件夹中供程序使用。在布局文件中添加一个文本框，用于显示手势名称；添加一个 GestureOverlayView 组件，用于绘制手势。注册监听，响应 Gesture 输入事件。在模拟器上绘制手势，用文本框显示对应的名称。

布局文件：

```xml
<RelativeLayout xmlns:android="http://schemas.android.com/apk/res/android"
    xmlns:tools="http://schemas.android.com/tools"
    android:layout_width="match_parent"
    android:layout_height="match_parent" >
    <TextView
        android:id="@+id/textView1"
        android:layout_width="wrap_content"
        android:layout_height="wrap_content"
        android:text="TextView" />
    <android.gesture.GestureOverlayView
        android:id="@+id/gestures"
        android:layout_below="@+id/textView1"
        android:gestureStrokeType="multiple"
        android:layout_height="fill_parent"
        android:layout_width="fill_parent">
    </android.gesture.GestureOverlayView>
```

```
</RelativeLayout>
```

逻辑代码如下：

```java
public class Gesture1Activity extends Activity {
    //显示手势名称
    private TextView view ;
    //声明Gesture库
    private GestureLibrary gLibrary;
    //Gesture输入的透明性重叠层
    private GestureOverlayView gestures;
    @Override
    public void onCreate(Bundle savedInstanceState) {
        super.onCreate(savedInstanceState);
        setContentView(R.layout.activity_gesture1);

        view = (TextView)findViewById(R.id.textView1);
        gestures = (GestureOverlayView) findViewById(R.id.gestures);
        //获取手势文件
        gLibrary = GestureLibraries.fromRawResource(this, R.raw.gestures);
        //载入Gestures,设置标题文本,提示是否载入成功
        if(gLibrary.load()){
            setTitle("手势文件装载成功。");
            }else{
            setTitle("手势文件装载失败。");
        }
        //注册监听,响应Gesture绘制事件
        gestures.addOnGesturePerformedListener(new OnGesturePerformedListener(){
            public void onGesturePerformed(GestureOverlayView overlay,
            Gesture gesture) {
                // TODO Auto-generated method stub
                //识别用户绘制的手势，是否在手势库中
                ArrayList predictions = gLibrary.recognize(gesture);
                //有可能匹配的手势
                if(predictions.size()>0){
                    //开始扫描手势
                    for(int i=0;i<predictions.size();i++){
                    //predictions保存了所有与当前手势可能匹配的候选手势
                    //获取手势库中的手势
                      Prediction prediction = (Prediction) predictions.get(i);
                      //相似度大于1
                    if (prediction.score>1.0) {
                        //显示手势名称
                        view.setText(prediction.name);
                        }
                    }
                }
            }
        });
    }
}
```

运行程序，在模拟器上绘制手势，文本框显示对应的名称，如图 13.3 所示。

图 13.3 输入法手势效果图

13.3　Wi-Fi

　　Wi-Fi 全称 Wireless Fidelity，是一种短程无线传输技术，能够在数百英尺范围内支持互联网接入的无线电信号。个人电脑、手持设备（如掌上电脑 PAD、手机）等终端可以通过这种无线方式互相连接，是当今使用最广的一种无线网络传输技术。Android 系统提供了 Wi-Fi 包（android.wifi）用于 Wi-Fi 应用，表 13-4 是对该包中重要类的说明。

表 13-4　Wi-Fi 包中重要的类

类　名	说　明
WifiManager	Wi-Fi 连接管理器，用于管理 Wi-Fi 连接
WifiInfo	描述了 Wi-Fi 连接状态
WifiConfiguration	代表已配置的 Wi-Fi 网络的配置信息

　　【示例 13-3】　获取 Wi-Fi 信息。新建项目 Wifi，获取 Wi-Fi 系统服务。在布局文件中添加两个 TextView 控件，一个用于显示获取的 Wi-Fi 信息，另一个用于显示 Wi-Fi 网络配置信息。

　　逻辑代码如下：

```
public class WifiActivity extends Activity {
    //Wi-Fi 管理器
    private WifiManager wifiManager;
    //Wi-Fi 信息
    private WifiInfo wifiInfo;
    //Wi-Fi 配置信息
    private List<WifiConfiguration> wifiConfigurations;
    private TextView tvWifiInfo,tvWifiConfigurations;
    @Override
    public void onCreate(Bundle savedInstanceState) {
        super.onCreate(savedInstanceState);
        setContentView(R.layout.activity_wifi);
        //获得 WifiManager 对象
        wifiManager = (WifiManager) getSystemService(Context.WIFI_SERVICE);
```

```java
        //获得连接信息对象
        wifiInfo = wifiManager.getConnectionInfo();
        tvWifiInfo = (TextView) findViewById(R.id.textView1);
        tvWifiConfigurations = (TextView) findViewById(R.id.textView2);
        //获得Wi-Fi信息
        StringBuffer sb = new StringBuffer();
            sb.append("Wifi信息\n");
            sb.append("MAC地址: " + wifiInfo.getMacAddress() + "\n");
            sb.append("接入点的BSSID: " + wifiInfo.getBSSID() + "\n");
            sb.append("IP地址(int):" + wifiInfo.getIpAddress() + "\n");
            sb.append("网络ID: " + wifiInfo.getNetworkId() + "\n");
            tvWifiInfo.setText(sb.toString());
        //得到配置好的网络
        wifiConfigurations = wifiManager.getConfiguredNetworks();
        tvWifiConfigurations.setText("已连接的无线网络\n");
        //得到配置信息
        for (WifiConfiguration wifiConfiguration : wifiConfigurations){
            tvWifiConfigurations.setText(tvWifiConfigurations.
            getText() + wifiConfiguration.SSID + "\n");
        }
    }
}
```

AndroidManifest.xml：

```xml
<manifest xmlns:android="http://schemas.android.com/apk/res/android"
    package="com.example.wifi"
    android:versionCode="1"
    android:versionName="1.0" >

    <uses-sdk
        android:minSdkVersion="8"
        android:targetSdkVersion="13" />
    <!--允许应用程序访问Wi-Fi的状态-->
    <uses-permission android:name="android.permission.ACCESS_WIFI_STATE"/>

    <application
        android:icon="@drawable/ic_launcher"
        android:label="@string/app_name"
        android:theme="@style/AppTheme" >
        <activity
            android:name=".WifiActivity"
            android:label="@string/title_activity_wifi" >
            <intent-filter>
                <action android:name="android.intent.action.MAIN" />

                <category android:name="android.intent.category.LAUNCHER" />
            </intent-filter>
        </activity>
    </application>

</manifest>
```

由于模拟器不支持Wi-Fi功能，所以本案例在支持Wi-Fi功能的真机设备上运行。首先要打开设备的Wi-Fi服务。在设置中，选择"无线和网络"，然后勾选WLAN连接到无线网络，如图13.4所示。

运行程序，获取 Wi-Fi 信息及其网络配置信息，效果如图 13.5 所示。

图 13.4 打开 Wi-Fi 服务

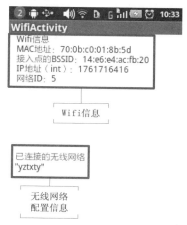

图 13.5 获取 Wi-Fi 信息

13.4 蓝牙 Bluetooth

蓝牙（Bluetooth）是一种支持设备短距离通信（一般 10m 内）的无线电技术。使用它，用户就能在包括移动电话、掌上电脑 PAD、无线耳机、笔记本电脑、相关外设等众多设备之间，进行无线信息交换。利用蓝牙技术，能够有效地简化移动通信终端设备之间的通信，也能够成功地简化设备与因特网 Internet 之间的通信。Android 系统提供了蓝牙包（android.bluetooth）用于蓝牙应用，表 13-5 是该包中重要的类和接口的说明。

表 13-5 蓝牙包重要类和接口

类/接口名	说明
BluetoothAdapter	代表本地蓝牙设备
BluetoothDevice	表示一个远程蓝牙设备
BluetoothServerSocket	用于侦听的蓝牙服务套接字
BluetoothSocket	已连接或正连接的蓝牙套接字
BluetoothClass	用于描述一个蓝牙设备的特性和性能

就某一次蓝牙通信而言，主动侦听连接的设备称为服务端，发起请求的设备称为客户端，无论客户端还是服务端，要想与其他设备进行蓝牙通信，通信双方都必须开启蓝牙服务。

【示例 13-4】获取服务端蓝牙设备信息。新建项目 Bluetooth，在布局文件中添加一个 TextView 控件，显示服务端蓝牙设备信息。

逻辑代码如下：

```
public class BluetoothActivity extends Activity {
    //请求码
    private static final int REQUEST_ENABLE_BT = 0;
    //本地蓝牙设备
    private BluetoothAdapter bAdapter;
```

```java
//显示服务端蓝牙设备信息
private TextView textView;

@Override
public void onCreate(Bundle savedInstanceState) {
    super.onCreate(savedInstanceState);
    setContentView(R.layout.activity_bluetooth);
    //获取本地蓝牙设备对象
    bAdapter = BluetoothAdapter.getDefaultAdapter();
    //如果设备为空
    if (bAdapter == null) {
    //使用Toast提示
     Toast.makeText(BluetoothActivity.this, "没有提供蓝牙通信", Toast.
     LENGTH_LONG).show();
      //调用finish()退出程序
        finish();
        return;
    }
    //如果设备不可用
    if (!bAdapter.isEnabled()) {
        //通过系统设置启动蓝牙
        Intent intent = new Intent(BluetoothAdapter.ACTION_
        REQUEST_ENABLE);
        startActivityForResult(intent, REQUEST_ENABLE_BT);
    }
    //查找已配对的设备
    Set<BluetoothDevice> pairedDevices = bAdapter.getBondedDevices();
    if (pairedDevices.size() > 0) {
        // Loop through paired devices
        //遍历已配对的设备信息
        for (BluetoothDevice device : pairedDevices) {
          textView = (TextView)findViewById(R.id.textView1);
          //显示已配对的设备信息
          textView.setText(device.getName() + "\n" + device.getAddress());
        }
    }
}
public void onActivityResult(int RequestCode, int ResultCode,Intent data){
    switch(RequestCode){
    case REQUEST_ENABLE_BT:
        //选择"是",Toast提示蓝牙已启动
        if(ResultCode == RESULT_OK){
            Toast.makeText(this.getApplicationContext(),"BT Launched!",
            Toast.LENGTH_SHORT).show();
        }
        //选择"否",Toast提示取消蓝牙启动
        else if(ResultCode == RESULT_CANCELED){
            Toast.makeText(this.getApplicationContext(),"Launched
            BT cancled!", Toast.LENGTH_SHORT).show();
        }
        break;
    }
}
}
```

AndroidManifest.xml：

```xml
<manifest xmlns:android="http://schemas.android.com/apk/res/android"
```

```xml
package="com.example.bluetooth"
android:versionCode="1"
android:versionName="1.0" >
<uses-sdk
    android:minSdkVersion="8"
    android:targetSdkVersion="13" />
<uses-permission android:name="android.permission.BLUETOOTH"/>
<uses-permission android:name="android.permission.BLUETOOTH_ADMIN"/>
<application
    android:icon="@drawable/ic_launcher"
    android:label="@string/app_name"
    android:theme="@style/AppTheme" >
    <activity
        android:name=".BluetoothActivity"
        android:label="@string/title_activity_bluetooth" >
        <intent-filter>
            <action android:name="android.intent.action.MAIN" />
            <category android:name="android.intent.category.LAUNCHER" />
        </intent-filter>
    </activity>
</application>
</manifest>
```

由于模拟器不支持蓝牙功能，所以本案例要在支持蓝牙功能的真机设备上测试。先运行一遍该程序，然后关闭（相当于将应用程序安装到手机中）。程序第一次启动后，如果手机没有打开蓝牙功能，则弹出"蓝牙权限请求"对话框，单击"是"按钮开启蓝牙功能，程序界面弹出 Toast 显示"BT Launched"，表示蓝牙服务已经启动，如图 13.6 所示。

启动了蓝牙服务后，开始扫描服务端设备，进行配对连接。配对成功后，运行程序，获取服务端蓝牙设备信息，如图 13.7 所示。

图 13.6　打开蓝牙服务　　　　　　图 13.7　获取服务端蓝牙设备信息

13.5　小　　结

本章主要讲解了 Android 中的手势和两种无线通信方式——Wi-Fi 和蓝牙的开发及应用。其中，触摸屏手势开发比较简单，比较容易掌握。输入法手势开发稍微复杂，需要读者多多练习，熟练掌握。无线通信方式在日常应用中比较广泛，希望读者结合相关 API 介绍，开发出具有更多功能的应用程序。

13.6 习　　题

1. 参考 13.2.2 节的内容，在 Gestures Builder 中创建手势。然后新建项目 Gesture，将手势文件导入到项目的 res 目录下的 raw 文件夹。如果该文件夹不存在，则需要新建该文件夹。在布局文件中添加 GestureOverlayView 控件用于绘制手势，然后显示相应的手势名称在文本框中，如图 13.8 所示。

图 13.8　输入法手势

【分析】本题目主要考查读者对输入法手势的掌握。首先需要手动创建手势，并为创建好的手势命名。然后将手势文件导入到项目中为项目所用。在程序中为 Gesture 注册监听，响应 Gesture 的输入事件。

【核心代码】本题的核心代码如下所示。

布局代码：

```
<android.gesture.GestureOverlayView
    android:id="@+id/gestures"
    android:layout_below="@+id/textView1"
    android:gestureStrokeType="multiple"
    android:layout_height="fill_parent"
    android:layout_width="fill_parent">
</android.gesture.GestureOverlayView>
```

逻辑代码：

```
gLibrary = GestureLibraries.fromRawResource(this, R.raw.gestures);
gestures.addOnGesturePerformedListener(new OnGesturePerformedListener(){
    public void onGesturePerformed(GestureOverlayView overlay,
    Gesture gesture) {
        // TODO Auto-generated method stub
        ArrayList predictions = gLibrary.recognize(gesture);
        if(predictions.size()>0){
            for(int i=0;i<predictions.size();i++){
                Prediction prediction = (Prediction) predictions.get(i);
                if (prediction.score>1.0) {
                    view.setText(prediction.name);
                }
```

```
            }
        }
    }
});
```

2. 使用支持 Wi-Fi 的真机设备，打开 Wi-Fi 服务，连接到无线网络。然后开发程序，获取 Wi-Fi 信息以及配置信息。

【分析】本题目主要考查读者对 android.wifi 包中重要类应用的掌握。首先要获取 Wi-Fi 系统服务，然后调用相关方法，获取信息。在开发过程中，需要添加用户权限"android.permission.ACCESS_WIFI_STATE"。可以参考 13.1 节的内容。

【核心代码】本题的核心代码如下所示。

```
wifiManager = (WifiManager) getSystemService(Context.WIFI_SERVICE);
wifiInfo = wifiManager.getConnectionInfo();
        StringBuffer sb = new StringBuffer();
        sb.append("Wifi 信息\n");
        sb.append("MAC 地址: " + wifiInfo.getMacAddress() + "\n");
        sb.append("接入点的 BSSID: " + wifiInfo.getBSSID() + "\n");
        sb.append("IP 地址(int):" + wifiInfo.getIpAddress() + "\n");
        sb.append("网络 ID: " + wifiInfo.getNetworkId() + "\n");
        tvWifiInfo.setText(sb.toString());
wifiConfigurations = wifiManager.getConfiguredNetworks();
tvWifiConfigurations.setText("已连接的无线网络\n");
for (WifiConfiguration wifiConfiguration : wifiConfigurations){
    tvWifiConfigurations.setText(tvWifiConfigurations.getText() +
        wifiConfiguration.SSID + "\n");
        }
```

3. 使用支持蓝牙功能的两台真机设备，分别打开蓝牙，两个设备配对连接。然后开发程序，获取服务端蓝牙设备名称及 MAC 地址。

【分析】本题目主要考查读者对 android.bluetooth 包中重要类应用的掌握。首先要获取本地蓝牙设备对象，然后调用相关方法获取已配对的设备信息。可以参考 13.3 节的内容。

【核心代码】本题的核心代码如下所示。

```
bAdapter = BluetoothAdapter.getDefaultAdapter();

if (!bAdapter.isEnabled()) {
    Intent intent = new Intent(BluetoothAdapter.ACTION_
    REQUEST_ENABLE);
    startActivityForResult(intent, REQUEST_ENABLE_BT);
}
Set<BluetoothDevice> pairedDevices = bAdapter.getBondedDevices();
if (pairedDevices.size() > 0) {
    // Loop through paired devices
    for (BluetoothDevice device : pairedDevices) {
        textView = (TextView)findViewById(R.id.textView1);
        textView.setText(device.getName() + "\n" + device.getAddress());
    }
}
public void onActivityResult(int RequestCode, int ResultCode,Intent data){
 switch(RequestCode)
 {
 case REQUEST_ENABLE_BT:
```

```
            if(ResultCode == RESULT_OK)
            {
                Toast.makeText(this.getApplicationContext(), "BT Launched!",
                Toast.LENGTH_SHORT).show();
            }
            else
                if(ResultCode == RESULT_CANCELED)
                {
                    Toast.makeText(this.getApplicationContext(),
                      "Launched BT cancled!",
                      Toast.LENGTH_SHORT).show();
                }
            break;
    }
}
```

第 14 章　Google 地图服务

提起 Google Maps（Google 地图），大家一定会想到其姊妹产品 Google Earth（Google 地球），全新的免费地图服务让 Google 在 2005 年震惊了整个互联网界。此后，各大门户网站纷纷推出了自己的地图服务，比如百度地图、8684 地图等。作为 Google 开发的主打产品，Andriod 提供了完备的地图服务功能。本章将结合实例开发，详细讲解 Google 地图的应用开发。

14.1　Google Maps

作为谷歌最为成功的一款网络服务之一，Google Maps 被广泛应用在旅游景点查询，以及线路导航等方面。

14.1.1　获取 Map API Key

Map API Key 是 Google Maps 的地图密钥。只有获取了 Map API Key，才能使用 Google Maps 服务。要获取 Map API Key，需要注册一个 Google 账号，该账号可在 Google 公司的官方网站免费申请。目前 Map API Key 已经升级为 Android Google Map API V2，本节将介绍获取 Map API Key 的过程。

1. 安装 Google Play services SDK

首先打开 Eclipse 中的 Window 菜单，单击 Android SDK Manager 选项，然后安装和更新 Extras 分类下的 Google Play services，如图 14.1 所示。

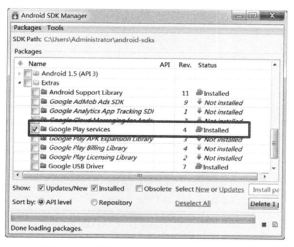

图 14.1　安装 Google Play services SDK

2. 获得证书指纹

要生成 API KEY，首先要找到 debug.keystore 文件，可以通过选择 Window |Preferences | Android |Build 命令来查看这个路径，比如本机中该文件的路径是 C:\Users\Administrator\.android\debug.keystore。然后通过终端执行如下命令，获取哈希码：

```
cd C:\Users\Administrator\.android\ debug.keystore
keytool -list -v -keystore debug.keystore
```

执行该命令后，就可以得到 MD5 及 SHA1 相关的字符串，如图 14.2 所示。

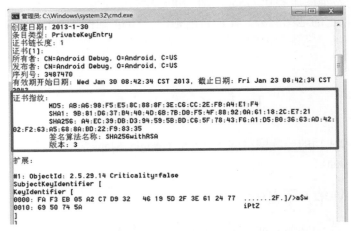

图 14.2　获得证书指纹

3. 生成 API KEY

得到证书指纹后，在浏览器地址栏中输入如下网址 https://code.google.com/apis/console/ 来获取 API KEY。进入页面后，先使用 Gmail 账号登录（如果没有需先申请），然后根据系统提示单击 Create project 创建 API 工程，然后会跳转到 services 页面，需要在这个页面打开 Google Maps Android API v2 选项（使之处于 on 的状态），如图 14.3 所示。

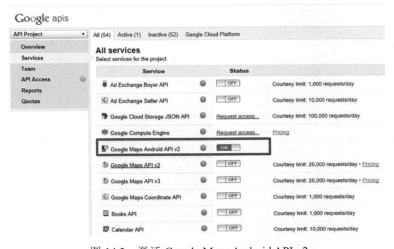

图 14.3　激活 Google Maps Android API v2

然后进入 API Access 页面，单击 Create new Android key…按钮，然后在对话框中填写 SHA1;com.package.name 这种形式的字符串。其中 SHA1 是签名生成的字符串，com.package.name 是我们所创建的工程名，具体示例如下：

```
0E:10:94:C7:B9:FD:15:62:27:FC:E7:FC:0C:05:59:A7:18:41:F5:B2;com.example.mapdemo
```

然后单击 Create 按钮，我们就得到了系统分配的 API KEY，如图 14.4 所示。

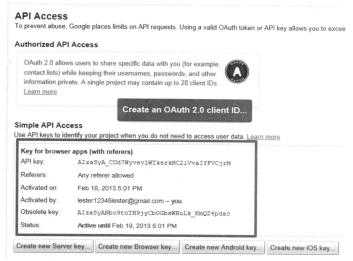

图 14.4　获取 API KEY

14.1.2　测试 Google Maps

获得了 Map API Key 之后，就可以对 Google 地图进行测试了。新建一个名为 mapdemo 的工程，然后对其中的配置文件进行修改和替换。

1. 修改 AndroidManifest.xml 文件

（1）在<application>元素之中加入如下子标签：

```xml
<meta-data android:name="com.google.android.maps.v2.API_KEY" android:value="your_api_key"/>
```

注意：在实际编程中，要替换上面的 your_api_key 为系统真正生成的那串 KEY 字符。
（2）在<application>标签之前添加许可信息（作为<manifest> 的子元素）：

```xml
<permission android:name="com.example.mapdemo.permission.MAPS_RECEIVE"
    android:protectionLevel="signature" />
    <uses-permission android:name="com.example.mapdemo.permission.MAPS_RECEIVE" />
    <uses-permission android:name="android.permission.INTERNET" />
    <uses-permission android:name="android.permission.WRITE_EXTERNAL_STORAGE" />
    <uses-permission android:name="android.permission.ACCESS_COARSE_LOCATION" />
    <uses-permission android:name="android.permission.ACCESS_FINE_LOCATION" />
    <uses-permission android:name="com.google.android.providers.gsf.permission.READ_GSERVICES" />
```

注意：代码中使用到了工程所创建的包名，也需要更改成自己的包名。

（3）在</application>标签之后添加对 OpenGL ES V2 特性支持（作为<manifest> 的子元素）：

```
<uses-feature android:glEsVersion="0x00020000" android:required="true" />
```

这样我们就完成了对文件 AndroidManifest.xml 的修改。

2. 修改布局文件 activity_main.xml

接着还需要对布局文件进行修改，来引入 Google 地图应用，具体代码如下所示。

```
<?xml version="1.0" encoding="utf-8"?>
<fragment xmlns:android="http://schemas.android.com/apk/res/android"
  android:id="@+id/map"
  android:layout_width="match_parent"
  android:layout_height="match_parent"
  class="com.google.android.gms.maps.MapFragment"/>
```

3. 替换 MainActivity.java 文件

最后需要替换 MainActivity.java 文件，具体代码如下所示。

```
package com.example.mapdemo;
import android.app.Activity;
import android.os.Bundle;
public class MainActivity extends Activity {
    @Override
    protected void onCreate(Bundle savedInstanceState) {
        super.onCreate(savedInstanceState);
        setContentView(R.layout.activity_main);
    }
}
```

这样我们就完成了 Google Map 应用的大部分工程，但是程序还不能正确运行，需要我们手动增加一个类库——Google Play services。

4. 添加 Google Play services 类库

在 Eclipse 里面依次选择 File | Import | Android | Existing Android Code Into Workspace 命令，然后单击 Next 按钮进入引进类库界面，单击 Browse...选项找到路径下的<android-sdk-folder>/extras/google/google_play_services/libproject/google-play-services_lib，然后单击 Finish 按钮。接着右击项目名，选择 properties 选项，在弹出的对话框中选择 Android 菜单，然后在下面的 Library 选项里面单击 Add 按钮，添加 google-play-services_lib。操作过程如图 14.5 所示。

5. 运行程序

至此，我们就完成了所有的配置过程，运行该项目，运行结果如图 14.6 所示。

14.1.3 Google Maps 相关类

在开发 Google Maps 服务时，会使用到 Google API 中的 com.google.android.map 包。重要的类如表 14-1 所示。

第 14 章　Google 地图服务

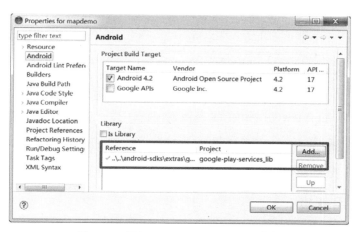

图 14.5　添加 Google Play services 类库　　　　图 14.6　程序运行结果

表 14-1　com.google.android.map 包中的重要类

类名称	类　说　明
MapActivity	用于显示 Google Map 的一个抽象 Activity 类，它需要连接底层网络，任何想要显示 MapView 的 Activity 都需要继承于 MapActivity，并且需要覆盖 isRouteDisplayer()方法
MapView	用于显示地图的 View 组件
MapController	用于控制和驱动地图的平移与缩放
OverLay	显示于地图之上的可绘制的对象
GeoPoint	这是一个包含经纬度位置的对象，单位是微度（度*1E-6）

其中，MapView 类包含了开发中常用的一些方法，这些方法如表 14-2 所示。MapController 类的相关方法如表 14-3 所示。

表 14-2　MapView 的相关方法

方 法 名 称	方 法 说 明
public MapController getController()	返回地图的 MapController
public final java.util.List<Overlay> getOverlays()	返回 Overlay 列表，这个列表中的任何一个 Overlay 都将被绘制
public int getZoomLevel()	返回当前地图的缩放级别
public Projection getProjection()	返回屏幕像素坐标与经纬度坐标之间的转换
public void setBuiltInZoomControls(boolean on)	设置是否启用内置的缩放控件。如果启用，MapView 将自动显示内置的缩放控件
public void displayZoomControls(boolean takeFocus)	显示缩放控件，可以选择是否请求焦点选中，以便通过按键访问
public void setSatellite(boolean on)	设置地图模式为"卫星"模式，装载带有道路名称的俯拍图像块，即打开卫星贴图

表 14-3　MapController 的相关方法

方 法 名 称	方 法 说 明
public void animateTo(GeoPoint point)	以给定的 point 开始动画显示地图
public voidsetZoom(int zoomLevel)	设置缩放级别

在开发中，还经常使用一个接口 Projection。该接口用于在屏幕像素 X/Y 坐标系和地球经纬度坐标系之间进行转换。Projectio.toPixels(GeoPoint in, Point out)方法，可以把给定的 GeoPoint 坐标变换到相对应的 MapView 左上角的屏幕像素坐标。其语法如下所示。

```
Projectio.toPixels( GeoPoint in , Point out )
```

其中，GeoPoint in 表示经纬度坐标，Point out 表示对应屏幕坐标。

14.1.4　Google Maps 应用开发

做好之前的准备工作，现在我们可以开发程序，使用 Google Maps 进行地图查询。其实现机制如图 14.7 所示。

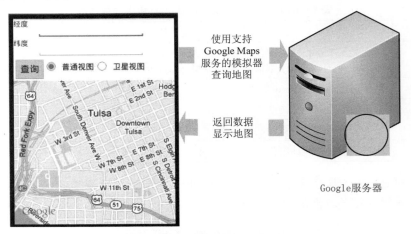

图 14.7　地图查询实现机制

【示例 14-1】　下面演示一个 Google Maps 地图查询应用。

1．新建项目

新建项目 MapQuery。创建时需注意，选择 Bulid SDK 为"GoogleAPIs (Google Inc.)-API Level 16"支持 Google Maps 服务，如图 14.8 所示。

2．界面设计

在布局界面添加两个文本框，分别显示"经度"和"纬度"；在每个文本框右侧添加编辑框，一个用于输入经度，一个用于输入纬度；添加"查询"按钮，单击按钮查询目的地；添加一个包含两个 RadioButton 的 RadioGroup，一个单击时显示普通视图；另一个单击时显示卫星视图；添加 MapView 组件显示地图。程序运行，初始界面如图 14.9 所示。

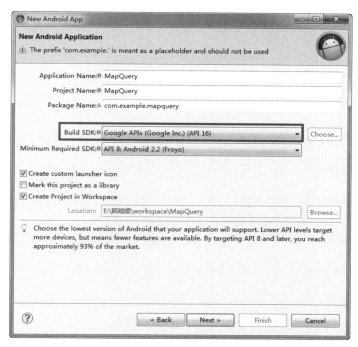

图 14.8　新建项目 MapQuery

图 14.9　程序初始界面图

布局文件：

```
<RelativeLayout
xmlns:android="http://schemas.android.com/apk/res/android"
    xmlns:tools="http://schemas.android.com/tools"
    android:layout_width="match_parent"
    android:layout_height="match_parent" >

    <TextView
        android:id="@+id/textView1"
        android:layout_width="wrap_content"
        android:layout_height="wrap_content"
        android:layout_alignParentLeft="true"
        android:layout_alignParentTop="true"
        android:text="经度" />

    <EditText
        android:id="@+id/editText1"
        android:layout_width="wrap_content"
        android:layout_height="wrap_content"
        android:layout_alignParentTop="true"
        android:layout_marginLeft="18dp"
        android:layout_toRightOf="@+id/textView1"
        android:ems="10"
        android:inputType="numberDecimal" >
    </EditText>

    <TextView
        android:id="@+id/textView2"
        android:layout_width="wrap_content"
        android:layout_height="wrap_content"
        android:layout_alignParentLeft="true"
```

```xml
        android:layout_below="@+id/editText1"
        android:text="纬度" />

    <EditText
        android:id="@+id/editText2"
        android:layout_width="wrap_content"
        android:layout_height="wrap_content"
        android:layout_alignLeft="@+id/editText1"
        android:layout_alignTop="@+id/textView2"
        android:ems="10"
        android:inputType="numberDecimal" />

    <Button
        android:id="@+id/button1"
        android:layout_width="wrap_content"
        android:layout_height="wrap_content"
        android:layout_alignParentLeft="true"
        android:layout_below="@+id/editText2"
        android:text="查询" />

    <RadioGroup
        android:id="@+id/radioGroup1"
        android:orientation="horizontal"
        android:layout_width="wrap_content"
        android:layout_height="wrap_content"
        android:layout_alignLeft="@+id/editText2"
        android:layout_marginLeft="10dp"
        android:layout_below="@+id/editText2" >

        <RadioButton
            android:id="@+id/radio0"
            android:layout_width="wrap_content"
            android:layout_height="wrap_content"
            android:layout_above="@+id/mapview1"
            android:layout_toLeftOf="@+id/radio1"
            android:checked="true"
            android:text="普通视图" />

        <RadioButton
            android:id="@+id/radio1"
            android:layout_width="wrap_content"
            android:layout_height="wrap_content"
            android:layout_above="@+id/mapview1"
            android:layout_alignParentRight="true"
            android:layout_marginRight="16dp"
            android:text="卫星视图" />
    </RadioGroup>
        <com.google.android.maps.MapView
            android:id="@+id/mapview1"
            android:layout_width="wrap_content"
            android:layout_height="wrap_content"
            android:layout_alignParentLeft="true"
            android:layout_alignParentRight="true"
            android:layout_below="@+id/button1"
            android:apiKey="0STtGbv8X9WzTBN3dFj9oDH9oiQXPL1aj7NDjlA"
            android:clickable="true"
            android:enabled="true" >
        </com.google.android.maps.MapView>
</RelativeLayout>
```

3. 代码实现

首先创建 SearchOverLay 类，用来在地图中指定的经纬度位置绘制一个搜索图标，标明该地点在地图中的确切位置。然后使主 Activity 继承于 MapActivity，显示地图视图。具体实现流程如图 14.10 所示。

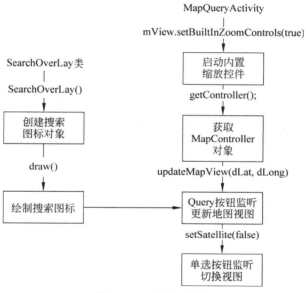

图 14.10　实现流程图

SearchOverLay：

```
//继承于 Overlay,用于绘制一个搜索图标,指示查询的位置
public class SearchOverLay extends Overlay {
    //搜索图标
    private Bitmap bitmap;
    //经纬度值对象
    private GeoPoint geoPoint;
    //搜索图标构造方法
    public SearchOverLay(Bitmap bitmap, GeoPoint geoPoint){
        this.bitmap = bitmap;
        this.geoPoint = geoPoint;
    }
    @Override
    //覆盖 draw 方法,绘制搜索图标到指定位置
    public void draw(Canvas canvas, MapView mView, boolean shadow) {
        if(!shadow){
            //获取 Projection 对象
            Projection projection = mView.getProjection();
            //屏幕坐标点
            Point point = new Point();
            //将真实地理坐标转化为屏幕上的坐标
            projection.toPixels(geoPoint, point);
            //绘制搜索图标
            canvas.drawBitmap(bitmap,
                point.x-bitmap.getWidth()/2,
                point.y-bitmap.getHeight(),
                null);
        }
```

            }
}

MapQueryActivity：

```java
//继承于MapActivity
public class MapQueryActivity extends MapActivity {
    //经纬度输入框
    private EditText latitude,longitude;
    //视图单选按钮,rb1显示普通视图,rb2显示卫星视图
    private RadioButton rb1,rb2;
    private RadioGroup rg;
    //搜索图标
    private Bitmap bitmap;
    //查询按钮
    private Button query;
    //声明MapView对象
    private MapView mView;
    //声明MapController对象
    private MapController controller;
    @Override
    public void onCreate(Bundle savedInstanceState) {
        super.onCreate(savedInstanceState);
         //设置无标题
         requestWindowFeature(Window.FEATURE_NO_TITLE);
         //全屏显示
         getWindow().setFlags(WindowManager.LayoutParams.FLAG_FULLSCREEN,
                    WindowManager.LayoutParams.FLAG_FULLSCREEN);
         //加载布局文件
        setContentView(R.layout.activity_map_query);
         //将drawable图片转换为bitmap格式
        bitmap = BitmapFactory.decodeResource(getResources(), R.drawable.
        ic_action_search);
        //获取MapView控件引用
        mView = (MapView)findViewById(R.id.mapview1);
        //启用内置缩放控件
        mView.setBuiltInZoomControls(true);
        //获取MapController对象
        controller = mView.getController();
        //设置缩放级别
        controller.setZoom(15);
         //查询按钮引用
        query = (Button)findViewById(R.id.button1);
        //查询按钮监听
        query.setOnClickListener(new OnClickListener() {
            public void onClick(View v) {
                // TODO Auto-generated method stub
                longitude = (EditText)findViewById(R.id.editText1);
                latitude = (EditText)findViewById(R.id.editText2);
                //获取用户输入的经度值
                String sLong = longitude.getText().toString().trim();
                //获取用户输入的纬度值
                String sLat = latitude.getText().toString().trim();
                //如果输入为空,则Toast提示重新输入
                if (sLong.equals("")||sLat.equals("")) {
                    Toast.makeText(MapQueryActivity.this,
                        "请重新输入",Toast.LENGTH_SHORT).show();
                }
```

```java
            //将经度值和纬度值转换为 double 类型
            double dLong = Double.parseDouble(sLong);
            double dLat =Double.parseDouble(sLat);
            //更新地图显示
            updateMapView(dLat, dLong);
        }
    });
    //单选按钮引用
    rb1 = (RadioButton)findViewById(R.id.radio0);
    rb2 = (RadioButton)findViewById(R.id.radio1);
    rg = (RadioGroup)findViewById(R.id.radioGroup1);
    //单选按钮监听
    rg.setOnCheckedChangeListener(new OnCheckedChangeListener() {
            public void onCheckedChanged(RadioGroup group, int checkedId) {
                // TODO Auto-generated method stub
                //单击 rb1 显示普通视图
                if (checkedId == rb1.getId()) {
                    mView.setSatellite(false);
                //单击 rb2 显示卫星视图
                }else if (checkedId == rb2.getId()) {
                    mView.setSatellite(true);
                }
            }
    });
}
@Override
//继承 MapActivity,覆盖该方法
protected boolean isRouteDisplayed() {
    // TODO Auto-generated method stub
    return false;
}
//自定义更新地图方法
private void updateMapView(double dLat, double dLong){
    //获取经纬度值
    GeoPoint point = new GeoPoint((int)(dLat*1E6), (int)(dLong*1E6));
    //显示缩放控件
    mView.displayZoomControls(true);
    //将地图移动到指定位置
    controller.animateTo(point);
    //返回 OverLay 列表
    List<Overlay> list = mView.getOverlays();
    //清空 list
    list.clear();
    //重绘搜索图标
    list.add(new SearchOverLay(bitmap,point));
}
```

AndroidManifest.xml：

```xml
<manifest xmlns:android="http://schemas.android.com/apk/res/android"
    package="com.example.mapquery"
    android:versionCode="1"
    android:versionName="1.0" >
    <uses-sdk
        android:minSdkVersion="8"
```

```xml
        android:targetSdkVersion="15" />
<uses-permission android:name="android.permission.INTERNET"/>
<application
    android:icon="@drawable/ic_launcher"
    android:label="@string/app_name"
    android:theme="@style/AppTheme" >
    <uses-library android:name="com.google.android.maps" />
    <activity
        android:name=".MapQueryActivity"
        android:label="@string/title_activity_map_query" >"
        <intent-filter>
            <action android:name="android.intent.action.MAIN" />
            <category android:name="android.intent.category.LAUNCHER" />
        </intent-filter>
    </activity>
</application>
</manifest>
```

注意：<user-library>必须为<application>的直接子类，否则程序报错。

运行程序，输入经纬度数据，单击"查询"按钮进行查询。"普通视图"和"卫星视图"切换显示对应的视图，效果如图 14.11 所示。

图 14.11　地图查询

14.2　Google Street View

Google Street View（谷歌街景）服务启动于 2007 年 5 月 25 日，由 Google 公司开发，是 Google 地图内的一项功能。谷歌街景由专用街景车进行拍摄，然后把 360 度实景拍摄照片放在谷歌地图里供用户使用，为用户提供了立体街道全景。

14.2.1　Google Street View 服务原理

Google Street View 街景服务的原理比较简单：当需要 Google 街景服务时，只需将包含经纬度信息的 Intent 启动内置的 com.google.android.street 应用程序即可。该 Intent 中包含信息如表 14-4 所示。

表 14-4 启动 Google 街景的 Intent 信息

字段名称	字 段 值	说 明
action	VIEW	查看
data	google.streetview:cbll=lat,lng&cbp=1,yaw,,pitch,zoom&mz=mapZoom	cbll 为必填参数，其中 lat 和 lng 分别代表纬度和经度。cbp 和 mz 为可选参数，代表了街景的立体视角及缩放尺寸

14.2.2 Google Street View 应用开发

下面通过具体案例，演示 Google Street View 的使用。

【示例 14-2】 新建项目 StreetView。创建支持 Google Maps 服务的模拟器。在布局界面添加两个文本框"经度"和"纬度"；添加两个输入框，一个用于输入经度；另一个用于输入纬度；添加 GO 按钮，单击时查询街景。实现流程如图 14.12 所示。

图 14.12 实现流程图

逻辑代码如下：

```java
public class StreetViewActivity extends Activity {
    //经纬度输入框
    private EditText ed1,ed2;
    //查询按钮
    private Button go;
    @Override
    public void onCreate(Bundle savedInstanceState) {
        super.onCreate(savedInstanceState);
        setContentView(R.layout.activity_street_view);
        go = (Button)findViewById(R.id.button1);
        //查询按钮监听
        go.setOnClickListener(new OnClickListener() {

            public void onClick(View v) {
                // TODO Auto-generated method stub
```

```
        ed1 = (EditText)findViewById(R.id.editText1);
        ed2 = (EditText)findViewById(R.id.editText2);
        //获取用户输入的经纬度值
        String sLong = ed1.getText().toString().trim();
        String sLat = ed2.getText().toString().trim();
        //如果输入为空,Toast 提示重新输入
        if (sLong.equals("")||sLat.equals("")) {
            Toast.makeText(StreetViewActivity.this,
            "请重新输入",Toast.LENGTH_SHORT).show();
            return;
        }
        //生成 Uri 字符串
        String sUri = "google.streetview:cbll=" + sLat + "," + sLong;
        //启动街景服务
        Intent intent = new Intent();
        intent.setAction(Intent.ACTION_VIEW);
        Uri uri = Uri.parse(sUri);
        intent.setData(uri);
        startActivity(intent);
        }
    });
    }
}
```

运行程序,输入经度值 0.73445,纬度值 51.3018(该经纬度定位到伦敦 Rawling Street)。单击 GO 按钮查询街景,效果如图 14.13 所示。

注意:由于 Google 街景服务并没有涵盖全球所有的地方,所以读者输入某些经纬度后单击 GO 按钮,有可能显示不出任何街景,如图 14.14 所示。

图 14.13　Google Street View 效果图

图 14.14　无街景显示

14.3　GPS 定位服务

GPS(Global Positioning System),全球定位系统的简称。在 Android 系统中,我们可以调用 android.location 类及其相关方法,使用移动设备提供的 GPS 定位服务获取位置信息。

14.3.1 GPS 相关类简介

GPS 定位服务的中心组件是 LocationManager 系统服务,它提供 API 来确定位置和方位。调用 getSystemService(Context.LOCATION_SERVICE)可获取 LocationManager 系统服务。获取 GPS 信息时,还会用到以下类或接口。

1. LocationManager 类

LocationManager 类可以获取系统的定位服务。这个服务允许应用程序定期获得 GPS 地理位置的更新数据,或者当设备进入或接近某一地理位置时,可以关闭应用程序的 Inent。LocationManager 类的相关方法如表 14-5 所示。

表 14-5 LocationManager 类的相关方法

方法名称	方法说明
List<String> getAllProviders()	获取所有与设备关联的定位模块的列表
String getBestProvider(Criteria criteria, boolean enabledOnly)	获取最能满足给定的标准(Criteria 对象)的提供者的名称
GpsStatus getGpsStatus(GpsStatus status)	获取 GPS 当前状态
Location getLastKnownLocation(String provider)	获取最近一次可用地点信息
public void requestLocationUpdates(String provider, long minTime, float minDistance, LocationListener listener)	添加一个 LocationListener 监听器
boolean isProviderEnabled(String provider)	判断参数所指的设备是否可用

2. Location 类

Location 类可以表示某一特定时间地理位置的相关信息,其方法如表 14-6 所示。

表 14-6 Location 类相关方法

方法名称	方法说明	方法名称	方法说明
getAccuracy()	获取精确度	getLongitude()	获取经度
getAltitude()	获取高度	getSpeed()	获取速度
getBearing()	获取方位	getTime()	获取时间
getLatitude()	获取纬度		

3. LocationProvider 类

LocationProvider 类是一个提供定位服务的抽象父类,定期报告移动设备所在地理位置的数据信息。

4. LocationListener 接口

当 GPS 位置有所改变时,LocationListener 接口用来接收来自 LoacationManager 的通知。这个 LocationListener 接口事先定义在 requestLocationUpdates(String,long.float, LocationListener)方法中。LocationListener 相关方法如表 14-7 所示。

5. Criteria 类

当需要为地理位置信息的获取设置查询条件时,需要创建一个 Criteria 对象。调用该对象的 set 方法设置查询条件，如表 14-8 所示。

表 14-7　LocationListener 相关方法

方 法 名 称	方 法 说 明
onLocationChanged(Location location)	GPS 位置信息被更新时调用该方法，根据 Location 参数可以读出 GPS 的详细信息
onProviderDisabled(String provider)	启动 Provider 时调用该方法
onProviderEnabled(String provider)	关闭 Provider 时调用该方法
onStatusChanged(String provider, int status, Bundle extras)	GPS 位置信息的状态被更新时调用该方法

表 14-8　Criteria 的 set 方法

方 法 名 称	方 法 说 明
public void setAccuracy(int accuracy)	设置精度
public void setAltitudeRequired(boolean altitudeRequired)	是否要求海拔高度
public void setBearingRequired(boolean bearingRequired)	是否要求方位信息
public void setPowerRequirement(int level)	电量要求
public void setSpeedRequired(boolean speedRequired)	是否要求速度

14.3.2　GPS 应用开发

开发程序获取 GPS 信息流程如图 14.15 所示。

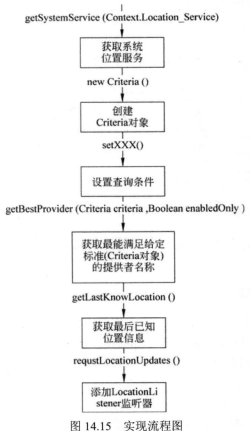

图 14.15　实现流程图

【示例 14-3】 使用 Location 及相关类获取 GPS 信息。新建一个项目 GPS。在界面添加 3 个文本框，分别用于显示获取的纬度、经度、方位信息。

逻辑代码如下：

```java
public class GPSActivity extends Activity {
    //声明位置管理器对象
    private LocationManager lManager;
    //显示 GPS 信息
    private TextView tv1,tv2,tv3;
    @Override
    public void onCreate(Bundle savedInstanceState) {
        super.onCreate(savedInstanceState);
        setContentView(R.layout.activity_gps);
        //位置监听
        LocationListener listener = new LocationListener() {
            public void onStatusChanged(String provider, int status, Bundle 
            extras){
                // TODO Auto-generated method stub
            }
            public void onProviderEnabled(String provider) {
                // TODO Auto-generated method stub
            }
            public void onProviderDisabled(String provider) {
                // TODO Auto-generated method stub
            }
            public void onLocationChanged(Location location) {
                // TODO Auto-generated method stub
                tv1 = (TextView)findViewById(R.id.textView1);
                tv2 = (TextView)findViewById(R.id.textView2);
                tv3 = (TextView)findViewById(R.id.textView3);
                //获取纬度信息显示
                tv1.setText("纬度: " + String.valueOf(location.
                getLatitude()));
                //获取经度信息显示
                tv2.setText("经度: " + String.valueOf(location.
                getLongitude()));
                //获取精确度显示
                tv3.setText("精确度: " + String.valueOf(location.
                getAccuracy()));
            }
        };
        //获取系统位置服务
        lManager = (LocationManager) getSystemService(Context.
        LOCATION_SERVICE);
        //获取 provider
        String bestProvider = lManager.getBestProvider(getCriteria(), true);
        //获取最后已知位置信息
        Location location =lManager.getLastKnownLocation(bestProvider);
        //添加位置监听
        lManager.requestLocationUpdates(
                            //添加位置监听
                            bestProvider,
                            //地理位置更新的最小时间间隔
                            0,
                            //位移变化的最短距离
                            0,
                            //监听器对象
```

```
            listener);
}
//创建一个 getCriteria()方法,返回 Criteria 对象,设置查询条件
private Criteria getCriteria() {
    // TODO Auto-generated method stub
    //创建 Criteria 对象
    Criteria criteria = new Criteria();
    //更好的位置精度要求
    criteria.setAccuracy(Criteria.ACCURACY_FINE);
    //低电量需求
    criteria.setPowerRequirement(Criteria.POWER_LOW);
    //返回 Criteria 对象
    return criteria;
    }
}
```

AndroidManifest.xml：

```xml
<manifest xmlns:android="http://schemas.android.com/apk/res/android"
    package="com.example.gps"
    android:versionCode="1"
    android:versionName="1.0" >

    <uses-sdk
        android:minSdkVersion="8"
        android:targetSdkVersion="15" />
    <uses-permission android:name="android.permission.ACCESS_FINE_LOCATION"/>
    <uses-permission android:name="android.permission.ACCESS_MOCK_LOCATION"/>
    <application
        android:icon="@drawable/ic_launcher"
        android:label="@string/app_name"
        android:theme="@style/AppTheme" >
        <activity
            android:name=".GPSActivity"
            android:label="@string/title_activity_gps" >
            <intent-filter>
                <action android:name="android.intent.action.MAIN" />

                <category android:name="android.intent.category.LAUNCHER" />
            </intent-filter>
        </activity>
    </application>

</manifest>
```

注意：如果是在真机测试该程序，则不需要添加用户权限"android.permission.ACCESS_MOCK_LOCATION"。

运行程序，启动模拟器。由于在模拟器中并没有提供和 GPS 设备直接连接的机制，所以模拟器界面并没有任何信息显示。因此我们需要在 DDMS 视图的 Emulator Control 面板中，找到 Location Control 窗体。手动输入经度、纬度，单击 Send 按钮，模拟器接收到 GPS 信息，如图 14.16 所示。

使用真机测试该程序，首先要启动设备的位置服务功能。在设置中，选择"位置服务"菜单，然后勾选"Google 的位置服务"选项，服务启动成功，如图 14.17 所示。

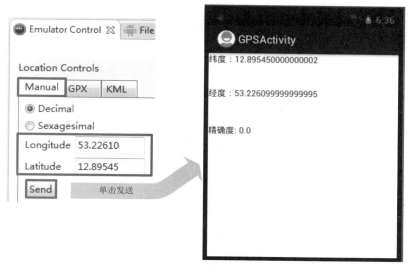

图 14.16　模拟器获取 GPS 信息

运行程序，通过位置服务获取当前所在位置的信息，效果如图 14.18 所示。

图 14.17　开启位置服务功能

图 14.18　真机获取 GPS 信息

14.4　小　　结

本章主要介绍了在 Android 系统中，如何获取地理位置，并应用于各种应用程序中。其中，GPS 和 Google Street View 的使用原理较为简单，比较容易掌握。Google Maps 相对较为复杂，需要读者多多练习，熟练掌握。

14.5 习　　题

1. 参考 14.1 节的内容，获取宿主电脑的 Map API Key。然后创建一个能够运行 Google 地图的模拟器，如图 14.19 所示。

图 14.19　能够运行 Google 地图的模拟器

【分析】本题目主要考查读者在使用 Google 地图之前，所做的准备工作的掌握。只有完成了本题，才可以使用 Android 系统提供的 Google Maps。

2. 新建一个项目 GPS，使用 Location 及相关类获取 GPS 信息。使用真机测试，获取读者所在地的经度和纬度等信息。

【分析】本题目主要考查读者对使用 Location 及相关类获取 GPS 信息的掌握。开发过程中，需要添加用户权限"android.permission.ACCESS_FINE_LOCATION"和"android.permission.ACCESS_MOCK_LOCATION"。可以参考 14.3 节的内容。

【核心代码】本题的核心代码如下所示。

```java
LocationListener listener = new LocationListener () {
            public void onStatusChanged(String provider, int
                status, Bundle extras) {
        // TODO Auto-generated method stub
    }
    public void onProviderEnabled(String provider) {
        // TODO Auto-generated method stub
    }
    public void onProviderDisabled(String provider) {
        // TODO Auto-generated method stub
    }
    public void onLocationChanged(Location location) {
        // TODO Auto-generated method stub
        tv1 = (TextView)findViewById(R.id.textView1);
        tv2 = (TextView)findViewById(R.id.textView2);
        tv3 = (TextView)findViewById(R.id.textView3);
        tv1.setText("纬度: " + String .valueOf(location.
        getLatitude()));
```

```java
            tv2.setText("经度: " + String.valueOf(location.
                getLongitude()));
            tv3.setText("精确度: " + String.valueOf(location.
                getAccuracy()));
        }
    };
    lManager = (LocationManager) getSystemService(Context.LOCATION_SERVICE);
    String bestProvider = lManager.getBestProvider(getCriteria(), true);
    Location location =lManager.getLastKnownLocation(bestProvider);
    lManager.requestLocationUpdates(bestProvider, 0, 0, listener);
}
private Criteria getCriteria() {
    // TODO Auto-generated method stub
    Criteria criteria = new Criteria();
    criteria.setAccuracy(Criteria.ACCURACY_FINE);
    criteria.setPowerRequirement(Criteria.POWER_LOW);
    return criteria;
}
```

第 15 章 Android 通信服务

不管是智能手机还是非智能手机，Android 作为手机操作系统家族的一员，在拨打电话、发送短信和 E-mail 邮件等手机通信方面有着非常出色的表现。本章将对 Android 手机通信功能的开发进行介绍。其中，包括短信的收发以及状态查询、电话的拨打与接听、来电的过滤和 E-mail 邮件的收发等。

15.1 电话控制

拨打电话是每台手机必备的功能，虽然在 Android 平台上可以通过程序实现各种让人目眩神迷的应用，但拨打电话这项最基本的功能，依然是每个 Android 工程师的必修课程。本节将对 Android 平台上与电话控制相关的知识进行介绍，包括拨打电话以及电话过滤等。

15.1.1 拨打电话

电话的拨打功能对于手机来说是必不可少的，而对于开发人员来说，掌握拨打电话的技术也是非常有必要的。本节将开发一个自定义的拨号程序替换系统自带的拨号程序，在读者掌握拨打电话技术的同时，了解如何用自己开发的应用程序替换系统自带的程序。

【示例 15-1】 开发一个自定义的拨号程序替换系统自带的拨号程序。案例的步骤如下所述。

（1）创建一个名为 Dial.java 的 Android 项目。
（2）准备图片资源，将应用程序所需要的图片资源存放到 res/drawable-mdpi 目录下。
（3）编写资源 XML，在图片目录的 res/drawable-mdpi 文件夹下创建图片的选择文件 myselector_del.xml，该文件用于设置删除按钮的背景图，其代码如下所示。

```xml
<?xml version="1.0" encoding="utf-8"?>        <!-- XML 的版本以及编码方式 -->
<selector
 xmlns:android="http://schemas.android.com/apk/res/android">
<!-- 定义一个 selector -->
  <item
   android:state_pressed="false"
   android:drawable="@drawable/del"/>          <!-- 添加状态为 false 的选项 -->
  <item
   android:state_pressed="true"
   android:drawable="@drawable/deldown"/>      <!--添加状态为 true 的选项-->
</selector>
```

注意：我们将该文件与图片资源存放到同一个目录下，然后在设置程序中 Button 的背景图片时指定该文件而不是某张图片，这样系统就会自动根据按钮的状态选择需要的图片充当按钮的背景图。

（4）继续编写其他按钮的选择文件，分别为删除按钮的背景 myselector_cancel.xml、拨号键的背景 myselector_dial.xml，以及数字的背景 myselector_num.xml。其代码与之前介绍过的 myselector_del.xml 文件基本相同，我们只以 myselector_cancel.xml 为例进行说明。

```xml
<?xml version="1.0" encoding="utf-8"?>
<selector
 xmlns:android="http://schemas.android.com/apk/res/android">
 <item
  android:state_pressed="false"
  android:drawable="@drawable/dialcancel"/>
 <item
  android:state_pressed="true"
  android:drawable="@drawable/dialcanceldown"/>
</selector>
```

（5）准备字符串资源，打开 Respectfully yours,/values 目录下的 strings.xml 文件，编写如下所示的代码，该文件定义了程序中需要的所有字符串资源。

```xml
<?xml version="1.0" encoding="utf-8"?>       <!-- XML 的版本以及编码方式 -->
    <resources>
        <string name="hello">Hello World, Dial!</string>
                                              <!-- 定义字符串 hello -->
        <string name="app_name">Dial</string>
                                              <!-- 定义字符串 app_name -->
        <string name="default_number">5556</string>
                                              <!--定义字符串 default_number -->
    </resources>
```

（6）编写颜色资源文件，在 res/values 目录下创建 colors.xml 文件，编写如下所示的代码，该文件将颜色资源统一定义到该处，以便程序的调试以及后期管理。

```xml
<?xml version="1.0" encoding="utf-8"?>       <!-- XML 的版本以及编码方式 -->
    <resources>
        <color name="red">#fd8d8d</color>     <!--定义 red 颜色 -->
        <color name="green">#9cfda3</color>   <!--定义 green 颜色-->
        <color name="blue">#8d9dfd</color>    <!--定义 blue 颜色-->
        <color name="white">#FFFFFF</color>   <!--定义 white 颜色-->
        <color name="black">#000000</color>   <!--定义 black 颜色-->
    </resources>
```

（7）搭建主页面。编写 main.xml 布局文件，该案例的布局如图 15.1 所示。

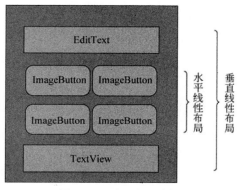

图 15.1　项目布局结构

在本文件中，我们首先定义一个垂直的线性布局，并依次向该布局添加一个水平的线性布局和一个垂直的线性布局，然后在水平布局中添加一个文本框和一个按钮控件，再向垂直线性布局添加若干个装有按钮控件的水平线性布局。main.xml 布局文件的主要代码如下所示。

```xml
<?xml version="1.0" encoding="utf-8"?>      <!-- XML 的版本以及编码方式 -->
<LinearLayout xmlns:android="http://schemas.android.com/apk/res/android"
    android:orientation="vertical"
    android:layout_width="fill_parent"
    android:layout_height="fill_parent">      <!-- 定义一个垂直的线性布局 -->
    <LinearLayout
        android:id="@+id/LinearLayout06"
        android:orientation="horizontal"
        android:layout_width="fill_parent"
        android:layout_height="wrap_content">
                                              <!-- 定义一个水平的线性布局-->
        <EditText
            android:text="@string/default_number"
            android:id="@+id/EditText01"
            android:layout_width="260dip"
            android:textSize="24dip"
            android:editable="false"
            android:enabled="false"
            android:singleLine="true"
            android:background="@color/white"
            android:textColor="@color/black"
            android:layout_marginRight="6dip"
            android:layout_marginLeft="10dip"
            android:layout_height="wrap_content"/>
                                              <!-- 添加一个 EditText 控件 -->
        <Button
            android:text=" "
            android:id="@+id/Button_del"
            android:textSize="24dip"
            android:layout_width="wrap_content"
            android:layout_height="wrap_content"
            android:background="@drawable/myselector_del"/> >
                                              <!-- 添加一个按钮控件 -->
    </LinearLayout>
    <LinearLayout
        android:id="@+id/LinearLayout01"
        android:orientation="vertical"
        android:layout_width="fill_parent"
        android:layout_height="wrap_content">
                                              <!-- 添加一个水平的线性布局 -->
        <LinearLayout
            android:id="@+id/LinearLayout02"
            android:orientation="horizontal"
            android:gravity="center_horizontal"
            android:layout_width="fill_parent"
            android:layout_height="wrap_content">
            ……
        </LinearLayout>
    </LinearLayout>
</LinearLayout>
```

(8) 编写主逻辑代码。在 Dial.java 文件中输入如下所示的代码。

```java
public class Dial extends Activity {
    int[] numButtonIds={                                   //数字按钮的 ID 数组
    R.id.Button00,R.id.Button01,R.id.Button02,
    R.id.Button03,R.id.Button04,R.id.Button05,
    R.id.Button06,R.id.Button07,R.id.Button08,
    R.id.Button09
    };
    public void onCreate(Bundle savedInstanceState) {
    super.onCreate(savedInstanceState);
        setContentView(R.layout.main);
        Button bDel=(Button)this.findViewById(R.id.Button_del);
        bDel.setOnClickListener(                           //为删除按钮添加监听器
            //OnClickListener 为 View 的内部接口,其实现者负责监听鼠标点击事件
            new View.OnClickListener(){
                public void onClick(View v){
                    EditText et=(EditText)findViewById(R.id.EditText01);
                    String num=et.getText().toString();
             num=(num.length()>1)?num.substring(0,num.length()-1):"";
                    et.setText(num);                       //组织字符串
                }
        });
        Button bDial=(Button)this.findViewById(R.id.Button_dial);
        bDial.setOnClickListener(                          //为拨号按钮添加监听器
            //OnClickListener 为 View 的内部接口,其实现者负责监听鼠标点击事件
            new View.OnClickListener(){
                public void onClick(View v){
                    //获取输入的电话号码
                    EditText et=(EditText)findViewById(R.id.EditText01);
                    String num=et.getText().toString();
                    //根据获取的电话号码创建 Intent 拨号
                    Intent dial = new Intent();
                    dial.setAction("android.intent.action.CALL");
                    dial.setData(Uri.parse("tel://"+num));
                    startActivity(dial);                   //激活打电话的 Activity
                }
            }
        );
        Button bCancel=(Button)this.findViewById(R.id.Button_cancel);
        bCancel.setOnClickListener(                        //为退出按钮添加监听器
            //OnClickListener 为 View 的内部接口,其实现者负责监听鼠标点击事件
         new View.OnClickListener(){
            public void onClick(View v){
                Dial.this.finish();                        //是否窗口
            }
         }
        );
        View.OnClickListener numListener=new View.OnClickListener(){
          public void onClick(View v){                     //为 0~9 数字按钮创建监听器
            Button tempb=(Button)v;                        //得到按钮的引用
            EditText et=(EditText)findViewById(R.id.EditText01);
                                                           //得到 EditText 的引用
            et.append(tempb.getText());                    //组织字符串
          }
        };
```

```
           for(int id:numButtonIds){                          //为所有数字按钮添加监听器
               Button tempb=(Button)this.findViewById(id);    //得到按钮
               tempb.setOnClickListener(numListener);         //添加监听
           }
       }
}
```

（9）为应用程序添加拨打电话的权限，并将该应用程序设置成系统的拨号程序，打开 AndroidManifest.xml 文件，其代码如下所示。

```
<?xml version="1.0" encoding="utf-8"?>
<manifest xmlns:android="http://schemas.android.com/apk/res/android"
    package="com.example.Dial" android:versionCode="1"
    android:versionName="1.0">
    <application                                    android:icon="@drawable/icon"
android:label="@string/app_name">
        <activity android:name="com.example.Dial.Dial"
                  android:label="@string/app_name">
            <intent-filter>
                <action android:name="android.intent.action.MAIN" />
                <category android:name="android.intent.category.LAUNCHER" />
            </intent-filter>
            <intent-filter>
<!-- 设置此应用程序也为系统拨号程序 -->
                <action android:name="android.intent.action.CALL_BUTTON" />
                <category android:name="android.intent.category.DEFAULT" />
            </intent-filter>
        </activity>
    </application>
    <uses-sdk android:minSdkVersion="7" />
       <!-- 设置此应用程序具有拨号权限 -->
       <uses-permission android:name="android.permission.CALL_PHONE"/>
</manifest>
```

此时启动两台模拟器，在其中一台模拟器（5554）中运行该程序并向另一台（5556）拨打电话，将观察到如图 15.2 所示的效果。第一次安装该程序后，以后每次用户再拨打电话时，会自动运行该程序来拨打电话，直到用户手动删除该程序。

图 15.2　程序运行效果

15.1.2　过滤电话

手机的电话过滤功能也是必不可少的，当用户不希望接到某些人的来电时，应该能够将这些电话号码添加到黑名单。在 Android 平台中，手机号码的过滤非常简单，只需通过对来电的号码进行检查,查看是否为不希望接听的号码,然后执行挂断、静音等操作。本节将对状态的查询方法进行介绍,并通过一个简单的案例来讲解如何快速地查询手机的状态,并在来电时对来电号码进行过滤。

【示例 15-2】 状态查询是通过添加 PhoneStateListener 监听实现的,接下来便逐步介绍该案例的开发过程。

(1) 首先创建一个名为 FilterPhone 的 Android 项目。

(2) 然后编写逻辑代码,打开 FilterPhone.java 文件,在方法中首先设置当前的用户界面,然后初始化 PhoneStateListener 监听器,并添加电话状态的监听。接着自定义 PhoneStateListener 监听类,该方法会在手机通话状态发生变化时被调用。该方法有两个入口参数,一个为手机的当前通话状态;另一个便为触发该方法的电话号码。当通话状态为来电状态时,判断来电号码是否为需要过滤的电话,判断出后便自行进行处理。其代码如下所示。

```java
public class FilterPhone extends Activity {
    String phoneNumber = "5556";                          //过滤的电话号码
    /** Called when the activity is first created. */
    @Override
    public void onCreate(Bundle savedInstanceState) {
                                                          //重写的onCreate()方法
        super.onCreate(savedInstanceState);
        setContentView(R.layout.main);                    //设置当前的用户界面
        MyPhoneStateListener myPhoneStateListene = new MyPhoneStateListener();
        TelephonyManager telManager = (TelephonyManager) getSystemService
         (TELEPHONY_SERVICE);
        telManager.listen(myPhoneStateListene, MyPhoneStateListener.
        LISTEN_CALL_STATE);
    }
    public class MyPhoneStateListener extends PhoneStateListener{
                                                          //创建监听器
        @Override
        public void onCallStateChanged(int state, String incomingNumber) {
            switch(state){
            case TelephonyManager.CALL_STATE_IDLE:        //待机状态
                Toast.makeText(FilterPhone.this, "当前手机为待机状态",
                Toast.LENGTH_LONG).show();
                break;
            case TelephonyManager.CALL_STATE_OFFHOOK:     //通话中
                Toast.makeText(FilterPhone.this, "当前手机为通话状态",
                Toast.LENGTH_LONG).show();
                break;
            case TelephonyManager.CALL_STATE_RINGING:     //来电状态
                if(incomingNumber.equals(phoneNumber)){
                                                          //是需要过滤的电话时
                    Toast.makeText(FilterPhone.this, "黑名单来电,可自行处理",
                    Toast.LENGTH_LONG).show();
                }
                else{                                     //不是需要过滤的电话时
                    Toast.makeText(FilterPhone.this, "当前手机为来电状态",
                    Toast.LENGTH_LONG).show();
                }
                break;
            }
            super.onCallStateChanged(state, incomingNumber);
```

```
        }
    }
}
```

注意：从 Android 1.0 版本开始，Google 便取消了挂断的方法，所以此处并没有做出任何处理代码，后面章节将介绍如何控制手机的铃声以及振动等功能，读者可以自行完善程序，使得当陌生人来电时将手机调成振动或者静音模式。

（3）为应用程序添加读取通话状态的权限，将下列代码添加到 AndroidManifest.xml 文件中的</manifest>标记之前即可。

```
<?xml version="1.0" encoding="utf-8"?>
<manifest xmlns:android="http://schemas.android.com/apk/res/android"
    package="com.FilterPhone"
    android:versionCode="1"
    android:versionName="1.0">
    <application android:icon="@drawable/icon" android:label="@string/app_name">
        <activity android:name="com.FilterPhone.FilterPhone"
            android:label="@string/app_name">
            <intent-filter>
                <action android:name="android.intent.action.MAIN" />
                <category android:name="android.intent.category.LAUNCHER" />
            </intent-filter>
        </activity>
    </application>
    <uses-sdk android:minSdkVersion="7" />
    <uses-permission android:name="android.permission.READ_PHONE_STATE"/>
</manifest>
```

运行该程序，当另一台模拟器向该模拟器拨打电话时，便会弹出 Toast 提示用户当前手机的通信状态，如图 15.3 所示。

图 15.3　程序运行效果

15.2　短信控制

短信的收发是手机不可缺少的一项重要功能，是使用频率最高的应用程序之一，所以掌握短信相关功能的开发是非常有必要的。本节将对 Android 平台下短信相关控制功能的开发进行详细介绍。

15.2.1 发送短信

本节将对 Android 平台的短信发送技术进行介绍，并通过一个开发短信发送功能的案例来详细介绍该技术的使用方法。需要注意的是，在带有短信收发功能的应用程序中，应该提前为应用程序指定短信权限，在发送时使用的是 PendingIntent 类，该类位于 android.app 包下。

【示例 15-3】 本例并不是简单地使用 Intent 激活 Android 自带的短信程序，而是自行开发的短信发送功能，我们甚至可以用自己开发的短信程序替换系统自带的短信程序。接下来详细介绍案例的开发步骤。

（1）首先创建一个名为 SendSMS 的 Android 项目。

（2）准备图片资源，将应用程序所需要的图片资源存放到 res/drawable-mdpi 目录下。

（3）准备字符串资源，打开 res/values 目录下的 strings.xml 文件，编写如下所示的代码，该文件定义了程序中需要的所有字符串资源。

```xml
<?xml version="1.0" encoding="utf-8"?>
<resources>
    <string name="app_name">SendSMS</string>
    <string name="dial">发送短信</string>
    <string name="sms">短信内容</string>
    <string name="tel">目标号码</string>
    <string name="telno">5556</string>
    <string name="smsnr">Hello</string>
</resources>
```

（4）编写颜色资源文件，在 res/values 目录下创建 colors.xml 文件，编写如下所示的代码，该文件将颜色资源统一定义到该处，以便程序的调试以及后期管理。

```xml
<?xml version="1.0" encoding="utf-8"?>
<resources>
    <color name="red">#fd8d8d</color>
    <color name="green">#9cfda3</color>
    <color name="blue">#8d9dfd</color>
    <color name="white">#FFFFFF</color>
    <color name="black">#000000</color>
    <color name="gray">#050505</color>
</resources>
```

（5）搭建用户界面，打开 main.xml，定义一个垂直的线性布局，然后依次向线性布局中添加 TextView、EditText 及 Button 控件，并分别为其进行参数的设置和指定 ID 值。其主要代码如下所示。

```xml
<?xml version="1.0" encoding="utf-8"?><!-- XML 的版本以及编码方式 -->
<LinearLayout xmlns:android="http://schemas.android.com/apk/res/android"
    android:orientation="vertical"
    android:layout_width="fill_parent"
    android:layout_height="fill_parent"
    android:background="@drawable/bbtc"
    android:gravity="bottom">            <!-- 添加一个垂直的线性布局 -->
    <TextView
```

```xml
      android:text="@string/tel"
      android:id="@+id/TextView02"
      android:textSize="20dip"
      android:textStyle="bold"
      android:textColor="@color/black"
      android:layout_width="wrap_content"
      android:layout_height="wrap_content"
      android:paddingLeft="5dip"/>              <!-- 添加一个TextView控件 -->
<EditText
      android:text="@string/telno"
      android:id="@+id/EditText02"
      android:layout_width="fill_parent"
      android:layout_height="wrap_content"/><!-- 添加一个EditText控件 -->
<TextView
      android:text="@string/sms"
      android:id="@+id/TextView01"
      android:layout_width="wrap_content"
      android:textSize="20dip"
      android:textStyle="bold"
      android:textColor="@color/black"
      android:paddingLeft="5dip"
      android:layout_height="wrap_content"/><!-- 添加一个TextView控件 -->
<EditText
      android:text="@string/smsnr"
      android:id="@+id/EditText01"
      android:layout_width="fill_parent"
      android:singleLine="false"
      android:gravity="top|left"
      android:layout_height="100dip"/>           <!-- 添加一个EditText控件 -->
<Button
      android:text="@string/dial"
      android:id="@+id/Button01"
      android:textSize="20dip"
      android:layout_width="fill_parent"
      android:layout_height="wrap_content"/>   <!-- 添加一个Button控件 -->
</LinearLayout>
```

（6）然后便进入主要逻辑代码的开发，在其中添加监听按钮，并自定义发送短信的方法，其主要代码如下所示。

```java
public class SendSMS extends Activity {
    /** Called when the activity is first created. */
    @Override
    public void onCreate(Bundle savedInstanceState) {
        super.onCreate(savedInstanceState);
        setContentView(R.layout.main);
        Button bdial=(Button)this.findViewById(R.id.Button01);
        bdial.setOnClickListener(                      // 为发送按钮添加监听器
            //OnClickListener为View的内部接口,其实现者负责监听鼠标点击事件
            new View.OnClickListener(){
                public void onClick(View v){
                    //获取输入的电话号码
                    EditText etTel=(EditText)findViewById(R.id.EditText02);
                    String telStr=etTel.getText().toString();
                    //获取输入的短信内容
                    EditText etSms=(EditText)findViewById(R.id.EditText01);
```

```java
            String smsStr=etSms.getText().toString();
            //判断号码字符串是否合法
            if(PhoneNumberUtils.isGlobalPhoneNumber
             (telStr)){                            //合法则发送短信
                v.setEnabled(false);
                                    //短信发送完成前将发送按钮设置为不可用
                sendSMS(telStr,smsStr,v);
            }
            else{                                  //不合法则提示
                Toast.makeText(
                    SendSMS.this,                  //上下文
                    "电话号码不符合格式!!! ",        //提示内容
                    5000                           //信息显示时间
                    ).show();
            }
        }
    });
}
//自己开发的直接发送短信的方法
private void sendSMS(String telNo,String smsStr,View v){
    PendingIntent pi=
        PendingIntent.getActivity(this, 0, new Intent(this,SendSMS.class), 0);
    SmsManager sms=SmsManager.getDefault();
    sms.sendTextMessage(telNo, null, smsStr, pi, null);
    //短信发送成功给予提示
    Toast.makeText(
        SendSMS.this,                              //上下文
        "恭喜你,短信发送成功! ",                    //提示内容
        5000                                       //信息显示时间
        ).show();
    v.setEnabled(true);              //短信发送完成后恢复发送按钮的可用状态
}
```

（7）为应用程序添加发送短信的权限。打开 AndroidManifest.xml 文件，在其中插入权限的声明代码。

```xml
<?xml version="1.0" encoding="utf-8"?>
<manifest xmlns:android="http://schemas.android.com/apk/res/android"
    package="com.SendSMS"
    android:versionCode="1"
    android:versionName="1.0">
  <application android:icon="@drawable/icon" android:label="@string/app_name">
      <activity android:name="com.SendSMS.SendSMS"
            android:label="@string/app_name">
          <intent-filter>
              <action android:name="android.intent.action.MAIN" />
              <category android:name="android.intent.category.LAUNCHER" />
          </intent-filter>
      </activity>
  </application>
  <uses-sdk android:minSdkVersion="7" />
  <!-- 设置此应用程序具有发短信权限 -->
  <uses-permission android:name="android.permission.SEND_SMS" />
</manifest>
```

启动两台模拟器，在一台模拟器中运行该程序，然后单击"发送短信"按钮，另一台模拟器便会接收到该信息，具体运行效果如图 15.4 所示。

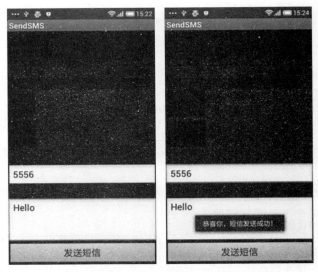

图 15.4　程序运行效果图

15.2.2　短信提示

上一节我们已经对 Android 平台下发送短信的技术进行了介绍，本节将对短信的接收方法进行介绍。接收到消息后，以 Toast 的形式提示用户短信的内容。

【示例 15-4】 案例的开发步骤如下所列。

（1）在 Eclipse 中新建一个名为 SMSRemind 的 Android 项目。

（2）准备字符串资源，打开 res/values 目录下的 strings.xml 文件，用下列代码替换其原有代码，定义程序中用到的字符串资源。

```xml
<?xml version="1.0" encoding="utf-8"?>
<resources>
    <string name="hello">Hello World, SMSRemind!</string>
    <string name="app_name"> SMSRemind</string>
    <string name="myString">恭喜您！程序安装成功！</string>
    <string name="ok">确定</string>
</resources>
```

（3）然后搭建用户界面，即编写布局文件 main.xml，该布局文件非常简单，只需向垂直的线性布局中依次添加一个 TextView，以及一个按钮控件。其代码如下所示。

```xml
<?xml version="1.0" encoding="utf-8"?>
<LinearLayout xmlns:android="http://schemas.android.com/apk/res/android"
    android:orientation="vertical"
    android:layout_width="fill_parent"
    android:layout_height="fill_parent"
    android:gravity="center_horizontal"
    >
    <TextView
        android:layout_width="fill_parent"
        android:layout_height="wrap_content"
```

```xml
        android:textSize="20px"
        android:gravity="center_horizontal"
        android:text="@string/myString"/>
    <Button
     android:id="@+id/myButton"
        android:layout_width="wrap_content"
        android:layout_height="wrap_content"
        android:textSize="20px"
        android:gravity="center_horizontal"
        android:text="@string/ok" />
</LinearLayout>
```

（4）然后开发 Activity 的实现类，打开 SMSRemind.java 文件，其代码如下所示。

```java
public class SMSRemind extends Activity {
    Button myButton;
    /** Called when the activity is first created. */
    @Override
    public void onCreate(Bundle savedInstanceState) {
        super.onCreate(savedInstanceState);
        setContentView(R.layout.main);
        myButton = (Button) findViewById(R.id.myButton);
        myButton.setOnClickListener(
          new OnClickListener(){
                @Override
                public void onClick(View v) {
                    // TODO Auto-generated method stub
                    System.exit(0);
                }
            }
        );
    }
}
```

（5）之后便进入本案例最主要的逻辑的开发。在目录下创建名为 MyBroadcastReceiver.java 的文件，开发如下所示的代码。

```java
public class MyBroadcastReceiver extends BroadcastReceiver{
    @Override
    public void onReceive(Context context, Intent intent) {
        // TODO Auto-generated method stub

        if(intent.getAction().equals("android.provider.Telephony.SMS_RECEIVED")){
            //收到的是短信
            Bundle bundle = intent.getExtras();
            if(bundle != null){
                Object[] myObject = (Object[])bundle.get("pdus");
                SmsMessage[] messages = new SmsMessage[myObject.length];
                for(int i=0; i<myObject.length; i++){
                    messages[i] = SmsMessage.createFromPdu((byte[])
                    myObject[i]);
                }
                StringBuilder sb = new StringBuilder();
                for(SmsMessage tempSmsMessage : messages){
                    sb.append("收到来自: \n");
                    sb.append(tempSmsMessage.getDisplayOriginatingAddress()+
                    "\n");
                    sb.append("内容为: \n");
                    sb.append(tempSmsMessage.getDisplayMessageBody());
                }
```

```
                    Toast.makeText(context, sb.toString(), Toast.LENGTH_
                        LONG).show();
                }
            }
        }
    }
}
```

（6）然后在 AndroidManifest.xml 中注册 BroadcastReceiver，并为应用程序添加接收短信的权限。

```xml
<?xml version="1.0" encoding="utf-8"?>
<manifest xmlns:android="http://schemas.android.com/apk/res/android"
    package="com.SMSRemind"
    android:versionCode="1"
    android:versionName="1.0">
    <application android:icon="@drawable/icon" android:label="@string/app_name">
        <activity android:name="com.SMSRemind.SMSRemind"
            android:label="@string/app_name">
            <intent-filter>
                <action android:name="android.intent.action.MAIN" />
                <category android:name="android.intent.category.LAUNCHER" />
            </intent-filter>
        </activity>
        <receiver android:name="com.SMSRemind.MyBroadcastReceiver">
            <intent-filter>
                <action android:name="android.provider.Telephony.SMS_RECEIVED"/>
            </intent-filter>
        </receiver>
    </application>
    <uses-sdk android:minSdkVersion="7" />
    <uses-permission android:name="android.permission.RECEIVE_SMS"/>
</manifest>
```

（7）在一台模拟器中运行该程序，然后单击"确定"按钮退出程序。此时通过另一台模拟器向安装该程序的模拟器发送短信息，安装该程序的模拟器便弹出 Toast 提示用户收到短信的信息。效果如图 15.5 所示。

图 15.5　程序运行效果图

15.2.3 短信群发

中国人好客，小到个人生日，大到过年过节，都会利用手机短信来表达祝福的心意。但是当我们要给多个朋友发送同一条短信时，如果从通讯录中一个一个添加联系人就会非常不便，这时我们可以使用短信群发功能来轻松实现一次发送一条短信给多人的效果。

【示例 15-5】 实现短信群发的效果。该案例的开发过程如下所列。

（1）创建一个名为 SMSMass 的 Android 项目。

（2）准备字符串资源，用下列代码替换 strings.xml 文件中原有代码，定义程序中用到的各个字符串资源。

```xml
<?xml version="1.0" encoding="utf-8"?>
<resources>
    <string name="hello">Hello World</string>
    <string name="app_name">短信群发</string>
    <string name="select">添加联系人</string>
    <string name="send">发送</string>
    <string name="people">您没有选取任何联系人</string>
</resources>
```

（3）搭建界面，打开布局文件 main.xml，设置外层的线性布局为垂直分布。接着向外层的线性布局中添加一个文本控件用来等待用户输入需要发送的短消息，并且为其指定 ID。开发如下所示的代码。

```xml
<?xml version="1.0" encoding="utf-8"?>
<LinearLayout xmlns:android="http://schemas.android.com/apk/res/android"
    android:orientation="vertical"
    android:layout_width="fill_parent"
    android:layout_height="fill_parent" >
    <EditText
        android:id="@+id/smsBody"
        android:layout_width="fill_parent"
        android:layout_height="wrap_content"
        android:text="@string/hello"/>
    <LinearLayout xmlns:android="http://schemas.android.com/apk/res/android"
        android:orientation="horizontal"
        android:layout_width="fill_parent"
        android:layout_height="wrap_content" >
        <Button
            android:id="@+id/select"
            android:layout_width="wrap_content"
            android:layout_height="wrap_content"
            android:text="@string/select"/>
        <Button
            android:id="@+id/send"
            android:layout_width="wrap_content"
            android:layout_height="wrap_content"
            android:text="@string/send"/>
    </LinearLayout>
    <EditText
        android:id="@+id/people"
        android:layout_width="fill_parent"
        android:layout_height="wrap_content"/>
</LinearLayout>
```

（4）然后开发 Activity 类，首先介绍按钮的事件响应部分的代码，在文件中创建一个 HashMap 容器，用于存放用户选择的联系人信息，包括姓名及电话号码。在 onCreate()方法中，设置当前显示的用户界面，然后得到 xml 文件中配置的各个控件的引用，并为两个按钮控件添加监听。本部分的代码如下所示。

```java
public class SMSMass extends Activity implements OnClickListener{
    Button select;                                          //选择联系人按钮
    Button send;                                            //发送
    EditText people;                                        //以及选择的联系人
    HashMap<String, String> peoples = new HashMap<String, String>();
                                                            //存储着算选择的所有
    @Override
    public void onCreate(Bundle savedInstanceState) {
        super.onCreate(savedInstanceState);
        setContentView(R.layout.main);                      //设置当前显示的用户界面
        select = (Button) this.findViewById(R.id.select);
                                                            //得到 select 按钮
        send = (Button) this.findViewById(R.id.send);       //得到 send 按钮
        people = (EditText) this.findViewById(R.id.people);
                                                            //得到 people 按钮
        select.setOnClickListener(this);                    //设置监听
        send.setOnClickListener(this);                      //设置监听
    }
    @Override
    public void onClick(View v) {                           //重写的按钮监听方法
        if(v == select){                                    //按下了选择联系人按钮
            Uri uri = Uri.parse("content://contacts/people");
            Intent intent = new Intent(Intent.ACTION_PICK, uri);
                                                            //创建 Intent
            startActivityForResult(intent, 1);              //切换到通讯录
        }
        else if(v == send){                                 //按下发送按钮
            v.setEnabled(false);                            //设置按钮为不可用
        //获取输入的短信内容
        EditText etSms=(EditText)findViewById(R.id.smsBody);
                                                            //得到 EditText 控件的引用
        String smsStr=etSms.getText().toString();           //得到短信的文本
            Set keySet = peoples.keySet();                  //得到键值集合
            Iterator ii = keySet.iterator();
            people.setText("");                             //置空
            while(ii.hasNext()){                            //循环
            Object key = ii.next();                         //得到键值
            String tempName = (String)key;                  //姓名
            String tempPhone = peoples.get(key);            //得到电话号码
            //判断号码字符串是否合法
            if(PhoneNumberUtils.isGlobalPhoneNumber(tempPhone)){
                                                            //合法则发送短信
                sendSMS(tempPhone,smsStr,v);                //发送短信
            }
         }
       }
     }
   private void sendSMS(String telNo,String smsStr,View v){
                                                            //自己开发的直接发送短信的方法
    PendingIntent pi=
```

```
            PendingIntent.getActivity(this, 0, new Intent(this,SMSMass.class), 0);
        SmsManager sms=SmsManager.getDefault();
        sms.sendTextMessage(telNo, null, smsStr, pi, null);    //发送短信
        v.setEnabled(true);                    //短信发送完成后恢复发送按钮的可用状态
    }
}
```

（5）接下来是对 onActivityResult()方法的完善。判断 requestCode 码与发送时是否相同，当相同时才需要处理，取得联系人的姓名和 ID，根据联系人的 ID 得到该条记录的电话号码。

```
protected void onActivityResult(int requestCode, int resultCode, Intent data){
    if(requestCode == 1){//requestCode 码与发送时相同时
        Uri myUri = data.getData();
        if(myUri != null){//当不为空时
            try{
                //得到 ContentResolver 对象
                ContentResolver cr = getContentResolver();
                Cursor c = managedQuery(myUri, null, null, null, null);
                c.moveToFirst();
                //取得联系人名字
                int nameFieldColumnIndex = c.getColumnIndex
                (PhoneLookup.DISPLAY_NAME);
                String sName = c.getString(nameFieldColumnIndex);
                //得到姓名
                //取得联系人 ID
                String contactId = c.getString(c.getColumnIndex
                (ContactsContract.Contacts._ID));
                Cursor phone = cr.query(ContactsContract.
                CommonDataKinds.Phone.CONTENT_URI, null, ContactsContract.
                CommonDataKinds.Phone.CONTACT_ID + " = " + contactId,
                null, null);
                //取得电话号码(当存在多个号码,只取一个)
                String strPhoneNumber = "";
                if(phone.moveToNext()){                //得到一个电话号码
                 strPhoneNumber = phone.getString
                 (phone.getColumnIndex(ContactsContract.
                 CommonDataKinds.Phone.NUMBER));
                }
                peoples.put(sName, strPhoneNumber);       //存放到容器中
                Set keySet = peoples.keySet();            //键值集合
                Iterator ii = keySet.iterator();
                people.setText("");                       //置空
                while(ii.hasNext()){
                 Object key = ii.next();                  //得到键值
                 String tempName = (String)key;           //姓名
                 String tempPhone = peoples.get(key);     //得到电话号码
                 people.setText(people.getText() + tempName + ":" +
                 tempPhone+"\n");
                }
            }catch(Exception e){                          //捕获异常
                e.printStackTrace();                      //打印异常信息
            }
        }
    }
    super.onActivityResult(requestCode, resultCode, data);
}
```

（6）为应用程序添加权限，将下列代码添加到 AndroidManifest.xml 文件中的</manifest>标签之前。

```xml
<?xml version="1.0" encoding="utf-8"?>
<manifest xmlns:android="http://schemas.android.com/apk/res/android"
    package="com.SMSMass"
      android:versionCode="1"
      android:versionName="1.0">
    <application android:icon="@drawable/icon" android:label="@string/app_name">
        <activity android:name="com.SMSMass.SMSMass"
                  android:label="@string/app_name">
            <intent-filter>
                <action android:name="android.intent.action.MAIN" />
                <category android:name="android.intent.category.LAUNCHER" />
            </intent-filter>
        </activity>

    </application>
    <uses-sdk android:minSdkVersion="7" />
     <uses-permission android:name="android.permission.READ_CONTACTS"/>
     <uses-permission android:name="android.permission.SEND_SMS"/>
</manifest>
```

运行该程序，效果如图 15.6 所示。当单击"添加联系人"按钮时，会显示手机的通讯簿，在通讯簿中选择某个联系人时，会自动回调该程序，并将选择的联系人添加到联系人列表中。当单击"发送"按钮时，会向联系人列表中的每个联系人发送一条短信息。

图 15.6　程序运行效果图

15.3　E-mail 控制

现在手机的功能越来越强大，很多用户都可以通过手机来办公。本节将介绍手机通信的另一个功能——E-mail 的发送。

SMTP（Simple Mail Transfer Protocol）即简单邮件传输协议，是 TCP/IP 协议族的一员，

它是一组用于由源地址到目的地址传送邮件的规则，由它来控制信件的中转发送方式，通过 SMTP 协议所指定的服务器便可将 E-mail 发送到收件人的邮箱。

通过 Android 平台发送 E-mail 是非常简单的，下面的代码为发送 E-mail 的核心代码。

```
String[] myReciver = new String[]{"Reciver"};    //寄件人
String[] mySubject = new String[]{"Subject"};    //主题
String myCc = "Cc";                              //副本
String myBody = "Body";                          //邮件内容
Intent myIntent = new Intent(android.content.Intent.ACTION_SEND);
                                                 //创建 Intent
myIntent.setType("plain/text");                  //设置邮件格式
myIntent.putExtra(android.content.Intent.EXTRA_EMAIL, myReciver);
                                                 //将寄件人放到 Intent 中
myIntent.putExtra(android.content.Intent.EXTRA_CC, myCc);
                                                 //将副本放到 Intent 中
myIntent.putExtra(android.content.Intent.EXTRA_SUBJECT, mySubject);
                                                 //将主图放到 Intent 中
myIntent.putExtra(android.content.Intent.EXTRA_TEXT, myBody);
                                                 //将邮件内容放到 Intent 中
startActivity(Intent.createChooser(myIntent, "标题"));
                                                 //打开 Gmail 发送邮件
```

Android 平台底层就是采用该协议进行通信的。实际上我们自己开发的邮件发送程序是通过调用 Android 内置的 Gmail 程序完成短信的发送的。需要注意的是，为了保证邮件能够正常的发送到指定地址的邮箱，E-mail 地址的格式必须是标准格式。

15.4 小　　结

本章主要对 Android 平台下的手机通信功能进行了介绍，包括短信的收发与状态查询、电话的拨打与过滤，以及 E-mail 的发送等。通过本章的学习，相信读者已经对 Android 平台下的手机通信功能的开发有了一定的了解。

15.5 习　　题

在触屏手机中，通常需要设置一个按钮来实现拨号处理。请使用 Intent 方式把电话号码传递给内置的拨号程序，然后内置拨号程序实现拨号处理操作。程序运行效果如图 15.7 所示。

【分析】利用 startActivity()方法将程序焦点交给内置的拨号程序，这样原来的 Activity 会成为失焦状态，并且还会发生 onPause 暂停事件，直到关闭拨号程序，焦点也交还给原来的 Activity。

【核心代码】编写文件 example.java，当用户单击图标按钮后，通过 android.intend.action.CALL_BUTTON 调用默认的拨号界面。主要代码如下所示。

```
public class example extends Activity
{
    private ImageButton myImageButton;
    @Override
```

图 15.7　屏幕触控拨打电话

```
public void onCreate(Bundle savedInstanceState)
{
  super.onCreate(savedInstanceState);
  setContentView(R.layout.main);
  myImageButton = (ImageButton) findViewById(R.id.myImageButton);
  myImageButton.setOnClickListener(new ImageButton.OnClickListener()
  {
    public void onClick(View v)
    {
      /* 调用拨号的画面 */
      Intent myIntentDial = new Intent("android.intent.action.CALL_BUTTON");
      startActivity(myIntentDial);
    }
  });
}
```

最后，需要在文件 AndroidManifest.xml 中声明 CALL_PHONE 权限，其主要代码如下所示。

```
<uses-permission android:name="android.permission.CALL_PHONE">
</uses-permission>
```

第 16 章　Android 特色应用开发

本章将要介绍的是 Android 手机特有 Feature 的开发,主要包括响应系统设置更改事件、设置手机外观和其他的特性。同时还将介绍如何在程序中获取 SIM 卡和电池电量等信息,最后以手机闹钟为例讲述如何开发特定功能的手机应用。

16.1　手机外观更改和提醒设置

本节将要介绍如何在程序中更改手机界面,如改变手机屏幕的壁纸;同时还将介绍如何设置手机的提醒设置,如振动、铃声大小等。在开发中合理地使用这些手机控制功能可以使应用程序提供更好的用户体验。

16.1.1　手机壁纸的改变

本节通过一个案例来说明如何在应用程序中对手机壁纸进行操作,包括获得手机壁纸、设置手机壁纸,以及还原手机壁纸到默认。

【示例 16-1】　该案例的开发步骤如下所述。

(1) 在 Eclipse 中新建一个项目 PhoneWallpaper,首先打开 res/values 目录下的 strings.xml,在<resources>和</resources>标记之间插入如下代码,声明的字符串资源将分别作为程序中的 3 个按钮的显示内容。

```xml
<?xml version="1.0" encoding="utf-8"?>
<resources>
    <string name="hello">Hello World, PhoneWallpaper!</string>
    <string name="app_name">PhoneWallpaper</string>
    <string name="getWall">获取当前墙纸</string>
    <string name="clearWall">恢复默认墙纸</string>
    <string name="setWall">设置为当前墙纸</string>
</resources>
```

(2) 打开项目 res/layout 目录下的 main.xml,在文件中声明一个垂直分布的线性布局,该布局中包括 3 个 Button 控件、1 个 Gallery 控件,以及 1 个 ImageView 控件。我们将其中已有代码替换为如下代码。

```xml
<?xml version="1.0" encoding="utf-8"?>
<LinearLayout xmlns:android="http://schemas.android.com/apk/res/android"
    android:orientation="vertical"
    android:layout_width="fill_parent"
    android:layout_height="fill_parent"
    >                                          <!-- 声明一个线性布局    -->
    <Button
     android:id="@+id/clearWall"
```

```xml
        android:layout_width="fill_parent"
        android:layout_height="wrap_content"
        android:text="@string/clearWall"
        />                                              <!-- 声明一个 Button     -->
    <ImageView
        android:id="@+id/currWall"
        android:layout_width="100px"
        android:layout_height="150px"
        android:layout_gravity="center_horizontal"
        />                                              <!-- 声明一个 ImageView   -->
    <Button
        android:id="@+id/getWall"
        android:layout_width="fill_parent"
        android:layout_height="wrap_content"
        android:text="@string/getWall"
        />                                              <!-- 声明一个 Button     -->
    <Gallery
        android:id="@+id/gallery"
        android:layout_width="fill_parent"
        android:layout_height="wrap_content"
        />                                              <!-- 声明一个 Gallery    -->
    <Button
        android:id="@+id/setWall"
        android:layout_width="fill_parent"
        android:layout_height="wrap_content"
        android:text="@string/setWall"
        />                                              <!-- 声明一个 Button     -->
</LinearLayout>
```

（3）下面来开发应用程序 Activity 部分的代码 PhoneWallpaper.java，首先来看其代码框架。

```java
public class PhoneWallpaper extends Activity {
    int [] imgIds ={                                //图片资源的 id 数组
        R.drawable.w1,
        R.drawable.w2,
        R.drawable.w3,
        R.drawable.w4
    };
    int selectedIndex = -1;                         //被选中的图片在 id 数组中的索引
    BaseAdapter ba = new BaseAdapter() {            //自定义的 BaseAdapter
        @Override
        public View getView(int position, View convertView, ViewGroup parent) {
            ImageView iv = new ImageView(PhoneWallpaper.this);
                                                    //新建一个 ImageView
            iv.setBackgroundResource(imgIds[position]);
                                                    //设置 ImageView 的背景图片
            iv.setScaleType(ImageView.ScaleType.CENTER_CROP);
            iv.setLayoutParams(new Gallery.LayoutParams(120, 120));
                                                    //设置相框中元素的大小
            return iv;
        }
        @Override
        public long getItemId(int arg0) {
            return 0;
        }
        @Override
        public Object getItem(int arg0) {
```

```
            return null;
        }
        @Override
        public int getCount() {
            return imgIds.length;
        }
    };
}
```

（4）接着来看 onCreate()方法的代码，该方法的主要功能是为程序中的各个按钮控件添加监听器。

```java
public void onCreate(Bundle savedInstanceState) {
    super.onCreate(savedInstanceState);
    setContentView(R.layout.main);                    //设置当前屏幕
    Button btnClearWall = (Button)findViewById(R.id.clearWall);
                                                      //获得 Button 对象
    btnClearWall.setOnClickListener(new View.OnClickListener() {
                                                      //添加 OnClickListener 监听器
        @Override
        public void onClick(View v) {                 //重写 onClick 方法
            try {
                PhoneWallpaper.this.clearWallpaper(); //还原手机壁纸
            } catch (IOException e) {                 //捕获并打印异常
                e.printStackTrace();
            }
        }
    });
    Button btnGetWall = (Button)findViewById(R.id.getWall);
                                                      //获得 Button 对象
    btnGetWall.setOnClickListener(new View.OnClickListener() {
                                                      //为 Button 添加 OnClickListener 监听器
        @Override
        public void onClick(View v) {
            ImageView iv = (ImageView)findViewById(R.id.currWall);
            iv.setBackgroundDrawable(getWallpaper());
                                                      //设置 ImageView 显示的内容为当前墙纸
        }
    });
    Gallery g = (Gallery)findViewById(R.id.gallery);
                                                      //获得 Gallery 对象
    g.setAdapter(ba);                                 //设置 Gallery 的 BaseAdapter
    g.setSpacing(5);                                  //设置每个元素之间的间距
    g.setOnItemClickListener(new OnItemClickListener() {
                                                      //为 Gallery 添加 OnItemClickListener 监听器
        @Override
        public void onItemClick(AdapterView<?> parent, View v, int
                position, long id) {
            selectedIndex = position;                 //记录被选中的图片索引
        }
    });
    Button btnSetWall = (Button)findViewById(R.id.setWall);
                                                      //获得 Button 对象
    btnSetWall.setOnClickListener(new View.OnClickListener() {
                                                      //为 Button 添加 OnClickListener 监听器
        @Override
        public void onClick(View v) {                 //重写 onClick 方法
            Resources r = PhoneWallpaper.this.getResources();
```

```
                                            //获得 Resources 对象
            InputStream in = r.openRawResource(imgIds[selectedIndex]);
                                            //获得 InputStream 对象
            try {
                setWallpaper(in);           //设置墙纸
            } catch (IOException e) {
                e.printStackTrace();
            }
        }
    });
}
```

（5）最后，还需要在应用程序的 AndroidManifest.xml 中为应用程序声明修改壁纸的权限，打开项目的 AndroidManifest.xml，在</manifest>标记之前输入如下代码。

```xml
<?xml version="1.0" encoding="utf-8"?>
<manifest xmlns:android="http://schemas.android.com/apk/res/android"
    package="com.PhoneWallpaper"
    android:versionCode="1"
    android:versionName="1.0">
  <application android:icon="@drawable/icon" android:label="@string/app_name">
      <activity android:name="com.PhoneWallpaper.PhoneWallpaper"
            android:label="@string/app_name">
          <intent-filter>
              <action android:name="android.intent.action.MAIN" />
              <category android:name="android.intent.category.LAUNCHER" />
          </intent-filter>
      </activity>

  </application>
    <uses-permission android:name="android.permission.SET_WALLPAPER" />

</manifest>
```

到此本案例的开发已经基本完成。运行本案例，用户可以在 Gallery 中选择图片，单击"设置为当前墙纸"按钮可以将壁纸设置为指定的图片，如图 16.1 所示。

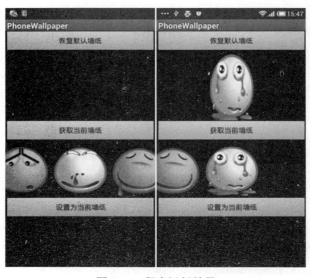

图 16.1　程序运行效果

16.1.2 手机振动的设置

本节将介绍如何在程序的代码中设置并启动手机振动。手机振动不仅可以作为来电的提醒，在应用程序中恰当地使用振动可以收到更好的效果。例如，在游戏中，当玩家失败一次，就进行一次振动提示。

在 Android 平台下不仅可以启动手机振动，还可以设置振动的周期、持续时间等详细参数。要想让手机启动振动，需要创建 Vibrator 对象，Vibrator 对象中常用的方法如表 16-1 所示。

表 16-1 Vibrator 对象常用方法及说明

方法名称	参数说明	方法说明
vibrate(long[] pattern, int repeat)	pattern：该数组中第一个元素是等待多长时间才启动振动，之后将会是开启和关闭振动的持续时间，单位为毫秒 repeat：重复振动时在 pattern 中的索引，如果设置为-1，则表示不重复振动	根据指定的模式进行振动
vibrate(long milliseconds)	milliseconds：振动持续的时间	启动振动，并持续指定的时间
cancel()		关闭振动

【示例 16-2】下面通过一个案例来说明如何在代码中获得 Vibrator 对象并调用指定的方法开启振动，该案例的开发步骤如下所述。

（1）在 Eclipse 中新建一个项目 PhoneVibrator，首先打开项目 res/values 目录下的 strings.xml，在<resources>和</resources>标记之间插入如下代码。

```xml
<?xml version="1.0" encoding="utf-8"?>
<resources>
    <string name="hello">Hello World, PhoneVibrator!</string>
    <string name="app_name">PhoneVibrator</string>
    <string name="vibrateOn">振动已启动</string>
    <string name="vibrateOff">振动已关闭</string>
    <string name="vibrate">启动振动</string>
    <string name="cancel">关闭振动</string>
</resources>
```

（2）打开项目 res/layout 目录下的 main.xml，在其中声明了一个垂直分布的线性布局，该线性布局中包含另外两个线性布局。其布局中声明了一个水平分布的线性布局，该布局中包含一个 ToggleButton 控件和一个 TextView 控件。

```xml
<?xml version="1.0" encoding="utf-8"?>
<LinearLayout xmlns:android="http://schemas.android.com/apk/res/android"
    android:orientation="vertical"
    android:layout_width="fill_parent"
    android:layout_height="fill_parent"
    >                                            <!-- 声明一个线性布局 -->
    <LinearLayout
        android:orientation="horizontal"
        android:layout_width="fill_parent"
        android:layout_height="wrap_content"
        >                                        <!-- 声明一个线性布局 -->
```

```xml
        <ToggleButton
            android:id="@+id/tb1"
            android:textOn="@string/cancel"
            android:textOff="@string/vibrate"
            android:checked="false"
            android:layout_width="wrap_content"
            android:layout_height="wrap_content"
            />                                      <!-- 声明一个 ToggleButton 控件 -->
        <TextView
            android:id="@+id/tv1"
            android:text="@string/vibrateOff"
            android:layout_width="wrap_content"
            android:layout_height="wrap_content"
            />                                      <!-- 声明一个 TextView -->
    </LinearLayout>
    <LinearLayout
        android:orientation="horizontal"
        android:layout_width="fill_parent"
        android:layout_height="wrap_content"
        >                                           <!-- 声明一个线性布局 -->
        <ToggleButton
            android:id="@+id/tb2"
            android:textOn="@string/cancel"
            android:textOff="@string/vibrate"
            android:checked="false"
            android:layout_width="wrap_content"
            android:layout_height="wrap_content"
            />                                      <!-- 声明一个 ToggleButton -->
        <TextView
            android:id="@+id/tv2"
            android:text="@string/vibrateOff"
            android:layout_width="wrap_content"
            android:layout_height="wrap_content"
            />                                      <!-- 声明一个 TextView -->
    </LinearLayout>
</LinearLayout>
```

（3）编写逻辑文件 PhoneVibrator.java，该文件声明了一个 Vibrator 对象的引用，该引用将会在 onCreate()方法中被赋值。onCreate()方法主要的功能是为 ToggleButton 控件添加 OnCheckedChangeListener 监听器。

```java
public class PhoneVibrator extends Activity {
    Vibrator vibrator;                              //声明一个 Vibrator 对象
    @Override
    public void onCreate(Bundle savedInstanceState) {
                                                    //重写 onCreate()方法
        super.onCreate(savedInstanceState);
        setContentView(R.layout.main);              //设置当前屏幕
        vibrator = (Vibrator)getSystemService(Service.VIBRATOR_SERVICE);
                                                    //创建 Vibrator 对象
        ToggleButton tb1 = (ToggleButton)findViewById(R.id.tb1);
                                                    //获得 ToggleButton 对象
        //设置 OnCheckedChangeListener 监听器
        tb1.setOnCheckedChangeListener(new OnCheckedChangeListener() {
            @Override
            //重写 onCheckedChanged 方法
            public void onCheckedChanged(CompoundButton buttonView,
            boolean isChecked){
```

```java
            if(isChecked){                    //判断ToggleButton的选中状态
                vibrator.vibrate(new long[]{1000,50,50,100,50}, -1);
                                              //启动振动
                TextView tv1 = (TextView)findViewById(R.id.tv1);
                                              //获得TextView
                tv1.setText(R.string.vibrateOn);
                                              //设置TextView控件内容
            }
            else{
                vibrator.cancel();            //关闭振动
                TextView tv1 = (TextView)findViewById(R.id.tv1);
                                              //获得TextView
                tv1.setText(R.string.vibrateOff);
                                              //设置TextView控件内容
            }
        }
    });
    ToggleButton tb2 = (ToggleButton)findViewById(R.id.tb2);
                                              //获得ToggleButton对象
    //设置OnCheckedChangeListener监听器
    tb2.setOnCheckedChangeListener(new OnCheckedChangeListener() {
        @Override
        //重写onCheckedChanged方法
        public void onCheckedChanged(CompoundButton buttonView,
            boolean isChecked) {
            if(isChecked){                    //判断ToggleButton的选中状态
                vibrator.vibrate(2500);       //启动振动
                TextView tv2 = (TextView)findViewById(R.id.tv2);
                                              //获得TextView
                tv2.setText(R.string.vibrateOn);
                                              //设置TextView控件内容
            }
            else{
                vibrator.cancel();            //关闭振动
                TextView tv2 = (TextView)findViewById(R.id.tv2);
                                              //获得TextView
                tv2.setText(R.string.vibrateOff);
                                              //设置TextView控件内容
            }
        }
    });
}
```

（4）最后还需要在应用程序的AndroidManifest.xml文件中声明振动的权限，打开项目AndroidManifest.xml，在</manifest>标记之前插入如下代码。

```xml
<?xml version="1.0" encoding="utf-8"?>
<manifest xmlns:android="http://schemas.android.com/apk/res/android"
    package="com.PhoneVibrator"
    android:versionCode="1"
    android:versionName="1.0">
    <application android:icon="@drawable/icon" android:label="@string/app_name">
        <activity android:name="com.PhoneVibrator.PhoneVibrator"
            android:label="@string/app_name">
            <intent-filter>
                <action android:name="android.intent.action.MAIN" />
```

```xml
            <category android:name="android.intent.category.LAUNCHER" />
        </intent-filter>
    </activity>
</application>

<uses-permission android:name="android.permission.VIBRATE" />
</manifest>
```

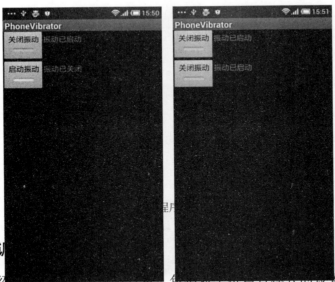

16.1.3 音量调节

本节将会介绍如何在程序中调整音量，包括对手机声音模式的设置和音量的调节。Android 对声音进行设置是通过 AudioManager 类来实现的，该类中包含了很多对声音模式和音量进行控制的方法。AudioManager 类的对象通过 Context 对象的 getSystemService(Context.AUDIO_SERVICE)来获得，其常用的对音量进行控制的方法见表 16-2 所示。

表 16-2 AudioManager类常用方法及说明

方 法 名 称	参 数 说 明	方 法 说 明
adjustStreamVolume(int streamType, int direction, int flags)	streamType：声音类型，可取的为 STREAM_ALARM、STREAM_DTMF、STREAM_MUSIC、STREAM_NOTIFICATION、STREAM_RING、STREAM_SYSTEM 和 STREAM_VOICE_CALL direction：调整音量的方向，可取的为 ADJUST_LOWER、ADJUST_RAISE 和 ADJUST_SAME flags：可选的标志位，可取的为 FLAG_ALLOW_RINGER_MODES、FLAG_PLAY_SOUND、FLAG_REMOVE_SOUND_AND_VIBRATE、FLAG_SHOW_UI 和 FLAG_VIBRATE	调整指定声音类型的音量
setMode(int mode)	mode：声音模式，可取的值为 NORMAL、RINGTONE 和 IN_CALL	设置声音模式
setRingerMode(int ringerMode)	ringerMode：铃声模式，可取的值为 RINGER_MODE_NORMAL、RINGER_MODE_SILENT 和 RINGER_MODE_VIBRATE	设置铃声模式
setStreamMute(int streamType, boolean state)	streamType：声音类型 state：是否使该类型声音静音的标志位	设置指定类型的声音是否需要静音

第 16 章 Android 特色应用开发

【示例 16-3】 下面通过一个案例来说明如何在代码中调节声音。在本案例中将会播放一段来自存储卡的音乐，用户可以在程序中使其静音或调整其音量大小，开发步骤如下所述。

（1）在 Eclipse 中新建一个项目 AdjustVolumn，在 res 目录下新建一个文件夹 raw，将程序中需要播放的声音文件 music.mp3 拷贝到该文件夹。

（2）打开 res/values 目录下的 strings.xml，在其\<resources>和\</resources>标记之间插入如下代码，声明的字符串资源将主要用做 Button 及 ToggleButton 控件的显示内容：

```xml
<?xml version="1.0" encoding="utf-8"?>
<resources>
    <string name="hello">Hello World, AdjustVolumn!</string>
    <string name="app_name">AdjustVolumn</string>
    <string name="btnPlay">播放音乐</string>
                                    <!-- 声明一个名为 btnPlay 的字符串 -->
    <string name="mute">静音</string>
                                    <!-- 声明一个名为 mute 的字符串 -->
    <string name="normal">正常</string>
                                    <!-- 声明一个名为 normal 的字符串 -->
    <string name="btnUpper">增大音量</string>
                                    <!-- 声明一个名为 btnUpper 的字符串 -->
    <string name="btnLower">减小音量</string>
                                    <!-- 声明一个名为 btnLower 的字符串 -->
</resources>
```

（3）打开 res/layout 目录下的 main.xml，在其中声明了一个垂直分布的线性布局，该布局中包括一个 Button 和另外一个线性布局。

```xml
<?xml version="1.0" encoding="utf-8"?>
<LinearLayout xmlns:android="http://schemas.android.com/apk/res/android"
    android:orientation="vertical"
    android:layout_width="fill_parent"
    android:layout_height="fill_parent"
    >                               <!-- 声明一个线性布局 -->
    <Button
        android:id="@+id/btnPlay"
        android:layout_width="fill_parent"
        android:layout_height="wrap_content"
        android:text="@string/btnPlay"
        />                          <!-- 声明一个 Button 控件 -->
    <LinearLayout
        android:orientation="horizontal"
        android:layout_width="wrap_content"
        android:layout_height="wrap_content"
        android:layout_gravity="center_horizontal"
        >                           <!-- 声明一个线性布局 -->
        <ToggleButton
            android:id="@+id/tbMute"
            android:layout_width="fill_parent"
            android:layout_height="wrap_content"
            android:textOn="@string/mute"
            android:textOff="@string/normal"
            android:layout_gravity="center_vertical"
            />                      <!-- 声明一个 ToggleButton 控件 -->
        <Button
            android:id="@+id/btnUpper"
```

```xml
            android:text="@string/btnUpper"
            android:layout_width="wrap_content"
            android:layout_height="wrap_content"
            />                                      <!-- 声明一个Button控件 -->
    <Button android:id="@+id/btnLower"
            android:text="@string/btnLower"
            android:layout_width="wrap_content"
            android:layout_height="wrap_content"
            />                                      <!-- 声明一个Button控件 -->
        </LinearLayout>
</LinearLayout>
```

（4）编写主逻辑文件 AdjustVolumn.java，在其中重写 onCreate()方法，该方法的主要功能是初始化成员变量并为布局文件中的 Button 及 ToggleButton 设置监听器，加载存储卡中的音乐文件并调用 MediaPlayer 的相关方法播放文件。

```java
public class AdjustVolumn extends Activity {
    MediaPlayer mp;                                 //声明MediaPlayer对象
    AudioManager am;                                //声明AudioManager对象
    @Override
    public void onCreate(Bundle savedInstanceState) {
        super.onCreate(savedInstanceState);
        setContentView(R.layout.main);              //设置当前屏幕
        am = (AudioManager)getSystemService(Service.AUDIO_SERVICE);
                                                    //创建AudioManager对象
        Button btnPlay = (Button)findViewById(R.id.btnPlay);
                                                    //获得Button对象
        btnPlay.setOnClickListener(new View.OnClickListener() {
                                                    //设置监听器
            @Override
            public void onClick(View v) {           //重写onClick方法
                try {
                    mp = MediaPlayer.create(AdjustVolumn.this, R.raw.music);
                    mp.setLooping(true);            //设置循环播放
                    mp.start();                     //播放声音
                }
                catch (Exception e) {               //捕获并打印异常
                    e.printStackTrace();
                }
            }
        });
        ToggleButton tbMute = (ToggleButton)findViewById(R.id.tbMute);
                                                    //获得ToggleButton对象
        tbMute.setOnCheckedChangeListener(new OnCheckedChangeListener() {
                                                    //添加监听器
            @Override
                                                    //重写onCheckedChanged方法
            public void onCheckedChanged(CompoundButton buttonView,
                boolean isChecked) {
                am.setStreamMute(AudioManager.STREAM_MUSIC, !isChecked);
                                                    //设置是否静音
            }
        });
        Button btnUpper = (Button)findViewById(R.id.btnUpper);
                                                    //获得Button对象
        btnUpper.setOnClickListener(new View.OnClickListener() {
                                                    //添加监听器
```

```java
            @Override
            public void onClick(View v) {                    //重写 onClick 方法
                am.adjustStreamVolume(AudioManager.STREAM_MUSIC, AudioManager.
ADJUST_RAISE,
                    AudioManager.FLAG_SHOW_UI);              //调高声音
            }
        });
        Button btnLower = (Button)findViewById(R.id.btnLower);
                                                              //获得 Button 对象
        btnLower.setOnClickListener(new View.OnClickListener() {
                                                              //添加监听器
            @Override
            public void onClick(View v) {                    //重写 onClick 方法
                am.adjustStreamVolume(AudioManager.STREAM_MUSIC,
AudioManager.ADJUST_LOWER, AudioManager.FLAG_SHOW_UI);  //调低声音
            }
        });
    }
}
```

完成上述步骤的开发之后,运行本案例,AdjustVolumn 的运行结果如图 16.3 所示。

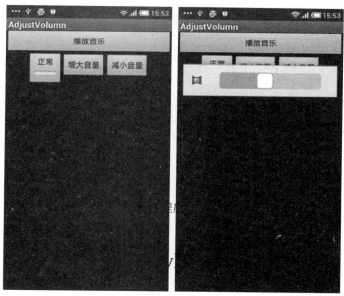

TelephonyManager 类位于 android.telephony 包下,主要提供了一系列用于访问与手机通讯相关的状态和信息的 get 方法。其中,包括手机 SIM 的状态和信息、电信网络的状态,以及手机用户的信息。本节将通过一个案例来说明如何从 TelephonyManager 对象中获取手机卡以及电信网络等信息。

【示例 16-4】 该案例的开发步骤如下所述。

(1) 在 Eclipse 中新建一个项目 PhoneManager,本程序中使用了多个字符串数组,而且这些数组在程序的运行过程中不会发生改变,为了方便管理,将这些数组集中声明在 XML 文件中。在 res/values 目录下新建一个文件 array.xml,在其中输入如下代码。

```xml
<?xml version="1.0" encoding="utf-8"?>
<resources>
```

```xml
<string-array name="listItem"><!-- 声明一个名为 listItem 的字符串数组 -->
    <item>设备编号</item>
    <item>SIM 卡国别</item>
    <item>SIM 卡序列号</item>
    <item>SIM 卡状态</item>
    <item>软件版本</item>
    <item>网络运营商代号</item>
    <item>网络运营商名称</item>
    <item>手机制式</item>
    <item>设备当前位置</item>
</string-array>
<string-array name="simState"><!-- 声明一个名为 simState 的字符串数组 -->
    <item>状态未知</item>
    <item>无 SIM 卡</item>
    <item>被 PIN 加锁</item>
    <item>被 PUK 加锁</item>
    <item>被 NetWork PIN 加锁</item>
    <item>已准备好</item>
</string-array>
<string-array name="phoneType"><!-- 声明一个名为 phoneType 的字符串数组 -->
    <item>未知</item>
    <item>GSM</item>
    <item>CDMA</item>
</string-array>
</resources>
```

（2）打开项目 res/layout 目录下的 main.xml，在其中声明一个垂直分布的线性布局，该布局中包含一个 ScrollView 控件。

```xml
<?xml version="1.0" encoding="utf-8"?>
<LinearLayout xmlns:android="http://schemas.android.com/apk/res/android"
    android:orientation="vertical"
    android:layout_width="fill_parent"
    android:layout_height="fill_parent"
    >                                          <!-- 声明一个线性布局 -->
    <ToggleButton
     android:id="@+id/tb"
     android:layout_width="fill_parent"
     android:layout_height="wrap_content"
     android:textOn="@string/on"
     android:textOff="@string/off"
     />                                        <!-- 声明一个 ToggleButton 控件 -->
    <TextView
         android:id="@+id/tv"
        android:layout_width="fill_parent"
        android:layout_height="wrap_content"
     />                                        <!-- 声明一个 TextView 控件 -->
</LinearLayout>
```

（3）编写逻辑文件 PhoneManager.java。该文件重写 BaseAdapter 对象的 getView 方法，该方法中首先创建一个线性布局 LinearLayout，该线性布局中主要包括两个 TextView，分别用于显示数据项的名称和数据项的值，如"网络运营商名称"为数据项的名称，"Android"为数据项的值。

```java
public class PhoneManager extends Activity {
    TelephonyManager tm;                       //声明 TelephonyManager 对象的引用
    String [] phoneType = null;                //声明表示手机制式的数组
    String [] simState = null;                 //声明表示 SIM 卡状态的数组
    String [] listItems = null;                //声明列表项的数组
    ArrayList<String> listValues = new ArrayList<String>();
```

```java
    BaseAdapter ba = new BaseAdapter() {
        @Override
        public View getView(int position, View convertView, ViewGroup parent)
{
            LinearLayout ll = new LinearLayout(PhoneManager.this);
            ll.setOrientation(LinearLayout.VERTICAL);
                                            //设置现象布局的分布方式
            TextView tvItem = new TextView(PhoneManager.this);
            TextView tvValue = new TextView(PhoneManager.this);
            tvItem.setTextSize(24);         //设置字体大小
            tvItem.setText(listItems[position]);   //设置显示的内容
            tvItem.setGravity(Gravity.LEFT);    //设置在父容器中的对齐方式
            ll.addView(tvItem);
            tvValue.setTextSize(18);            //设置字体大小
            tvValue.setText(listValues.get(position)); //设置显示的内容
            tvValue.setPadding(0, 0, 10, 10);       //设置四周边界
            tvValue.setGravity(Gravity.RIGHT);  //设置在父容器中的对齐方式
            ll.addView(tvValue);            //将TextView添加到线性布局中
            return ll;
        }
        @Override
        public long getItemId(int position) {   //重写getItemId方法
            return 0;
        }
        @Override
        public Object getItem(int position) {   //重写getItem方法
            return null;
        }
        @Override
        public int getCount() {                 //重写getCount方法
            return listItems.length;
        }
    };
```

(4)接着我们来看 onCreate()方法,该方法通过 getSystemService 方法创建了 TelephonyManager 对象。

```java
public void onCreate(Bundle savedInstanceState) {
    super.onCreate(savedInstanceState);
    setContentView(R.layout.main);
    tm = (TelephonyManager)getSystemService(Context.TELEPHONY_
    SERVICE);
    listItems = getResources().getStringArray(R.array.listItem);
                                        //获得XML文件中的数组
    simState = getResources().getStringArray(R.array.simState);
                                        //获得XML文件中的数组
    phoneType = getResources().getStringArray(R.array.phoneType);
                                        //获得XML文件中的数组
    initListValues();                   //初始化列表项的值
    ListView lv = (ListView)findViewById(R.id.lv);
                                        //获得ListView对象
    lv.setAdapter(ba);
}
```

(5)下面我们介绍 initListValues()方法,该方法的代码如下所示。

```java
public void initListValues(){                   //方法:获取各个数据项的值
    listValues.add(tm.getDeviceId());           //获取设备编号
    listValues.add(tm.getSimCountryIso());      //获取SIM卡国别
    listValues.add(tm.getSimSerialNumber());    //获取SIM卡序列号
```

```
        listValues.add(simState[tm.getSimState()]);        //获取 SIM 卡状态
                                                            //获取软件版本
        listValues.add((tm.getDeviceSoftwareVersion()==null?tm.getDeviceSoft
        wareVersion():"未知"));        listValues.add(tm.getNetworkOperator());
                                                            //获取网络运营商代号
        listValues.add(tm.getNetworkOperatorName());        //获取网络运营商名称
        listValues.add(phoneType[tm.getPhoneType()]);       //获取手机制式
        listValues.add(tm.getCellLocation().toString());    //获取设备当前位置
        }
```

注意：上述代码的主要功能是通过调用 TelephonyManager 不同的 get 方法获取手机 SIM 卡及电信网络的相关状态和信息。将这些数据值存放到 ArrayList 中以便 ListView 显示。

（6）由于访问 TelephonyManager 中的位置及手机状态信息需要相应的权限，所以还需要在应用程序的 AndroidManifest.xml 文件中声明权限。

```xml
<?xml version="1.0" encoding="utf-8"?>
<manifest xmlns:android="http://schemas.android.com/apk/res/android"
    package="com.PhoneManager"
    android:versionCode="1"
    android:versionName="1.0">
    <application android:icon="@drawable/icon" android:label="@string/app_name">
        <activity android:name="com.PhoneManager.PhoneManager"
            android:label="@string/app_name">
            <intent-filter>
                <action android:name="android.intent.action.MAIN" />
                <category android:name="android.intent.category.LAUNCHER" />
            </intent-filter>
        </activity>
    </application>
    <uses-permission android:name="android.permission.ACCESS_COARSE_LOCATION"/>
    <uses-permission android:name="android.permission.READ_PHONE_STATE"/>
</manifest>
```

完成上述步骤的开发之后，运行本案例，项目 PhoneManager 的运行效果如图 16.4 所示。

图 16.4　程序运行效果图

16.3 手机电池电量

手机电池电量的获取在应用程序的开发中也很常用，Android 系统中的手机电池电量发生变化的消息是通过 Intent 广播来实现的，常用的 Intent 的 Action 有 ACTION_BATTERY_CHANGED、ACTION_BATTERY_LOW 和 ACTION_BATTERY_OKAY。

当我们想要在程序中获取电池电量的信息时，需要为应用程序注册 BroadcastReceiver 组件，当特定的 Action 事件发生时，系统将会发出相应的广播，应用程序就可以接收广播并进行相应的处理。

【示例 16-5】 本节将会通过一个案例来说明如何在代码中获取手机电池的电量，本案例中的 BroadcastReceiver 组件用于捕获 ACTION_BATTERY_CHANGED 动作，其开发步骤如下所述。

（1）在 Eclipse 中新建一个项目 Battery，首先打开项目 res/values 目录下的 strings.xml，在其中的<resources>和</resources>标记之间插入如下代码，声明的字符串资源将作为程序中的 ToggleButton 控件显示的内容。

```xml
<?xml version="1.0" encoding="utf-8"?>
<resources>
    <string name="hello">Hello World, Battery!</string>
    <string name="app_name">Battery</string>
    <string name="on">停止获取电量信息</string><!-- 声明名为 on 的字符串资源 -->
    <string name="off">获取电量信息</string>   <!-- 声明名为 off 的字符串资源 -->
</resources>
```

（2）打开项目 res/layout 目录下的 main.xml，将其中已有代码替换为如下代码，在其中声明了一个垂直分布的线性布局，该布局中包括一个 ToggleButton 控件和一个 TextView 控件。

```xml
<?xml version="1.0" encoding="utf-8"?>
<LinearLayout xmlns:android="http://schemas.android.com/apk/res/android"
    android:orientation="vertical"
    android:layout_width="fill_parent"
    android:layout_height="fill_parent"
    >                                       <!-- 声明一个线性布局 -->
    <ToggleButton
     android:id="@+id/tb"
     android:layout_width="fill_parent"
     android:layout_height="wrap_content"
     android:textOn="@string/on"
     android:textOff="@string/off"
     />                                     <!-- 声明一个 ToggleButton 控件 -->
    <TextView
        android:id="@+id/tv"
        android:layout_width="fill_parent"
        android:layout_height="wrap_content"
        />                                  <!-- 声明一个 TextView 控件 -->
</LinearLayout>
```

（3）编写逻辑文件 Battery.java，在其中声明了 MyBatteryReceiver 对象的引用，MyBatteryReceiver 类继承自 BroadcastReceiver 类，该类的主要功能是接收系统发出的电池电量改变的广播。

```java
public class Battery extends Activity {
```

```java
    MyBatteryReceiver mbr = null;
    @Override
    public void onCreate(Bundle savedInstanceState) {
                                                    //重写 onCreate 方法
        super.onCreate(savedInstanceState);
        setContentView(R.layout.main);
        mbr = new MyBatteryReceiver();          //创建 Broadcast 组件对象
        ToggleButton tb = (ToggleButton)findViewById(R.id.tb);
                                                //获得 ToggleButton 对象
        tb.setOnCheckedChangeListener(new OnCheckedChangeListener() {
                                                //设置监听器
            @Override
            public void onCheckedChanged(CompoundButton buttonView,
            boolean isChecked){
                if(isChecked){
                    IntentFilter filter = new IntentFilter(Intent.ACTION
                    _BATTERY_CHANGED);
                    registerReceiver(mbr, filter);
                                                //注册 BroadcastReceiver
                }
                else{
                    unregisterReceiver(mbr);
                                                //取消注册的 BroadcastReceiver
                    TextView tv = (TextView)findViewById(R.id.tv);
                    tv.setText(null);           //清空 TextView 中显示的内容
                }
            }
        });
    }
    private class MyBatteryReceiver extends BroadcastReceiver{
        @Override
        public void onReceive(Context context, Intent intent) {
                                                //重写 onReceiver()方法
            int current = intent.getExtras().getInt("level");
                                                //获得当前电量
            int total = intent.getExtras().getInt("scale");
                                                //获得总电量
            int percent = current*100/total;    //计算百分比
            TextView tv = (TextView)findViewById(R.id.tv);
                                                //获得 TextView 对象
            tv.setText("现在的电量是："+percent+"%。");
                                                //设置 TextView 显示的内容
        }
    }
}
```

完成了上述步骤的开发后，运行本案例，项目 Battery 的运行效果如图 16.5 所示。

图 16.5　程序运行效果

16.4 手机闹钟

AlarmManager 类提供了访问系统定时服务的途径，开发人员可以在程序中设置某个应用程序在未来的某个时刻被执行。当 AlarmManager 定时时间到了之后，当初注册的 Intent 对象将会被系统广播，进而启动目标程序。

在程序运行，需要使用 AlarmManager 时，可以通过 Context 对象的 getSystemService (Context.ALARM_SERVICE)方法来获得 AlarmManager 对象。

【示例 16-6】 本节将通过一个案例来说明 AlarmManager 的用法，在程序中，可以在时间选择对话框设置闹钟的时间，在设置的时间到了的时候，会调用指定的 Activity。该案例的开发步骤如下所述。

（1）在 Eclipse 中新建一个项目 PhoneAlarm。首先打开项目 res/values 目录下的 strings.xml，在<resources>和</resources>标记之间插入如下代码，声明的字符串资源将会作为按钮以及对话框中各个部分显示的内容。

```xml
<?xml version="1.0" encoding="utf-8"?>
<resources>
    <string name="hello">Hello World, PhoneAlarm!</string>
    <string name="app_name">PhoneAlarm</string>
    <string name="btn">设置闹钟</string>
    <string name="alarmTitle">闹钟</string>
    <string name="alarmMsg">时间到了！</string>
    <string name="alarmButton">知道了</string>
</resources>
```

（2）打开项目 res/layout 目录下的 main.xml，在其中声明了一个垂直分布的线性布局，该布局中包含一个 Button 控件。

```xml
<?xml version="1.0" encoding="utf-8"?>
<LinearLayout xmlns:android="http://schemas.android.com/apk/res/android"
    android:orientation="vertical"
    android:layout_width="fill_parent"
    android:layout_height="fill_parent"
    >                                               <!-- 声明一个线性布局 -->
    <Button
        android:id="@+id/btn"
        android:text="@string/btn"
        android:layout_width="fill_parent"
        android:layout_height="wrap_content"
        />                                          <!-- 声明一个Button控件 -->
</LinearLayout>
```

（3）编写逻辑文件 PhoneAlarm.java，在其中创建了一个 Calendar，该 Calendar 中记录了当前的系统时间。

```java
public class PhoneAlarm extends Activity {
    Calendar c = Calendar.getInstance();
    final int DIALOG_TIME = 0;                      //设置对话框id
    AlarmManager am;                                //声明AlarmManager对象
    @Override
```

```java
public void onCreate(Bundle savedInstanceState) {
    super.onCreate(savedInstanceState);
    setContentView(R.layout.main);                    //设置当前屏幕
    am = (AlarmManager)getSystemService(Context.ALARM_SERVICE);
                                                       //创建 AlarmManager 对象
    Button btn = (Button)findViewById(R.id.btn);
                                                       //获得 Button 对象
    btn.setOnClickListener(new View.OnClickListener() {    //设置监听器
        @Override
        public void onClick(View v) {
                                                       //重写 onClick 方法
            showDialog(DIALOG_TIME);                   //显示时间选择对话框
        }
    });
}
```

（4）本步骤将对 PhoneAlarm.java 文件中的 onCreateDialog()方法进行详细的介绍，其代码如下所示。

```java
protected Dialog onCreateDialog(int id) {              //重写 onCreateDialog 方法
    Dialog dialog = null;
    switch(id){                                        //对 id 进行判断
    case DIALOG_TIME:
        dialog=new TimePickerDialog(                   //创建 TimePickerDialog 对象
            this,
            new TimePickerDialog.OnTimeSetListener(){
                                                       //创建 OnTimeSetListener 监听器
                @Override
                public void onTimeSet(TimePicker tp, int hourOfDay, int minute) {
                    Calendar c=Calendar.getInstance();
                                                       //获取日期对象
                    c.setTimeInMillis(System.currentTimeMillis());
                                                       //设置 Calendar 对象
                    c.set(Calendar.HOUR, hourOfDay);
                                                       //设置闹钟小时数
                    c.set(Calendar.MINUTE, minute);
                                                       //设置闹钟的分钟数
                    c.set(Calendar.SECOND, 0);
                                                       //设置闹钟的秒数
                    c.set(Calendar.MILLISECOND, 0);
                                                       //设置闹钟的毫秒数
                    //创建 Intent 对象
                    Intent intent = new Intent(PhoneAlarm.this,
                    AlarmReceiver.class);
                    //创建 PendingIntent
                    PendingIntent pi = PendingIntent.getBroadcast
                    (PhoneAlarm.this, 0, intent, 0);
                    am.set(AlarmManager.RTC_WAKEUP, c.getTimeInMillis(),
                    pi);                               //设置闹钟
                    Toast.makeText(PhoneAlarm.this, "闹钟设置成功",
                        Toast.LENGTH_LONG).show();     //提示用户
                }
            },
            c.get(Calendar.HOUR_OF_DAY),               //传入当前小时数
            c.get(Calendar.MINUTE),                    //传入当前分钟数
            false
```

```
        );
        break;
default:break;
}
        return dialog;
```

（5）下面进行 AlarmReceiver.java 的开发，AlarmReceiver 继承自 Broadcast 类，其主要的功能是接收闹钟时间到后被广播的 Intent 对象。其代码如下所示。

```
public class AlarmReceiver extends BroadcastReceiver{
    @Override
    public void onReceive(Context context, Intent intent) {
                                            //重写 onReceive 方法
        Intent i = new Intent(context,AlarmActivity.class);
                                            //创建 Intent 对象
        i.addFlags(Intent.FLAG_ACTIVITY_NEW_TASK);
                                            //设置标志位
        context.startActivity(i);           //启动 Activity
    }
}
```

说明：AlarmReceiver 类的代码比较简单，它的主要功能是通过重写父类的 onReceive 方法实现的。闹钟时间到了之后将会向该类发送 Intent，其 onReceive 方法主要进行的工作是创建另外一个 Intent 并根据该 Intent 启动 Activity。

（6）最后来介绍 AlarmActivity 类的开发。AlarmActivity 类是闹钟时间到了之后显示给用户的提醒界面，其代码如下所示。

```
public class AlarmActivity extends Activity{
    @Override
    protected void onCreate(Bundle savedInstanceState) {
        super.onCreate(savedInstanceState);
        new AlertDialog.Builder(AlarmActivity.this)
        .setTitle(R.string.alarmTitle)              //设置标题
        .setMessage(R.string.alarmMsg)              //设置内容
        .setPositiveButton(                         //设置按钮
            R.string.alarmButton,
            new OnClickListener() {                 //为按钮添加监听器
                @Override
                public void onClick(DialogInterface dialog, int which){
                    AlarmActivity.this.finish();
                                            //调用 finish 方法关闭 Activity
                }
            })
        .create().show();                           //显示对话框
    }
}
```

说明：在 AlarmActivity 的 onCreate 方法中，主要进行的工作是创建一个 AlertDialog 并将其显示到屏幕。该 AlertDialog 中主要包含一些提示信息和关闭按钮。

（7）完成功能代码的开发之后，还需要在 AndroidManifest.xml 中声明自定义的 BroadcastReceiver 组件和 Activity 类，打开项目的 AndroidManifest.xml 文件，在

</application>标记之前插入如下代码。

```xml
<?xml version="1.0" encoding="utf-8"?>
<manifest xmlns:android="http://schemas.android.com/apk/res/android"
    package="com.PhoneAlarm"
    android:versionCode="1"
    android:versionName="1.0">
    <application android:icon="@drawable/icon" android:label="@string/app_name">
        <activity android:name="com.PhoneAlarm.PhoneAlarm"
              android:label="@string/app_name">
            <intent-filter>
                <action android:name="android.intent.action.MAIN" />
                <category android:name="android.intent.category.LAUNCHER" />
            </intent-filter>
        </activity>
        <activity android:name="com.PhoneAlarm.AlarmActivity"/>
        <receiver android:name="com.PhoneAlarm.AlarmReceiver" android:process=":remote"/>
    </application>
</manifest>
```

完成了上述的开发之后,运行本案例,程序运行之后首先单击"设置闹钟"按钮,在弹出的时间选择对话框中确定闹钟的时间,如图 16.6 所示。

图 16.6 程序运行效果

16.5 小　　结

本章主要介绍如何在程序的开发过程中对手机的特性进行设置,主要包括手机外观设置、提醒设置,以及系统的设置等内容。掌握了这些设置手机特性的方法可以使开发出来的程序界面更加友好,功能更加合理。

16.6 习　　题

请开发一个闹钟程序,可以设置闹钟时间,也可以取消闹钟设置。当到了设置时间时,弹出对话框提示闹钟时间,程序运行效果如图 16.7 所示。

图 16.7　闹钟的实现

【分析】开发该应用程序，获取系统闹铃服务，实现闹铃设置和取消功能。新建 AlarmRecriver，监听系统时间，当到了闹铃时间，弹出对话框提示。

【核心代码】新建项目，在布局文件中，添加一个 TextView 控件显示设置信息；添加两个 Button 控件，一个用于设置闹铃；另一个用于取消设置；在 MainActivity 中，为两个按钮添加监听，获取系统闹铃服务，实现闹铃设置和取消功能；新建 AlarmActivity 显示闹钟对话框；新建 AlarmRecriver，监听系统时间，当到了闹铃时间，弹出对话框提示。在 AndroidManifest.xml 中注册，使用 AlarmActivity 和 AlarmRecriver。

（1）在 MainActivity 中，为两个按钮添加监听，获取系统闹铃服务，实现闹铃设置和取消功能，代码如下。

```java
public class MainActivity extends Activity {
    TextView textView;
    Button set,cancle;
    Calendar calendar;
    @Override
    protected void onCreate(Bundle savedInstanceState) {
        super.onCreate(savedInstanceState);
        setContentView(R.layout.activity_main);
        textView = (TextView)findViewById(R.id.textView1);
        set = (Button)findViewById(R.id.button1);
        cancle = (Button)findViewById(R.id.button2);
        //设置按钮监听
        set.setOnClickListener(new OnClickListener() {
```

```java
            @Override
            public void onClick(View v) {
                // TODO Auto-generated method stub
                //日历对象
                calendar = Calendar.getInstance();
                //当前系统时间
                calendar.setTimeInMillis(System.currentTimeMillis());
                int hour = calendar.get(Calendar.HOUR_OF_DAY);
                int minute = calendar.get(Calendar.MINUTE);
                //弹出时间选择对话框设置时间
                new TimePickerDialog(MainActivity.this,
                        new TimePickerDialog.OnTimeSetListener() {
                            @Override
                            public void onTimeSet(TimePicker view, int
                            hourOfDay, int minute) {
                                // TODO Auto-generated method stub
                                //设置后的时间
                                calendar.setTimeInMillis(System.
                                currentTimeMillis());
                                calendar.set(Calendar.HOUR_OF_DAY,
                                hourOfDay);
                                calendar.set(Calendar.MINUTE, minute);
                                calendar.set(Calendar.SECOND, 0);
                                calendar.set(Calendar.MILLISECOND, 0);
                                //到了闹钟设置时间,就运行AlarmReceiver
                                Intent intent = new Intent
                                (MainActivity.this, AlarmReceiver.class);
                                //创建PendingIntent
                                PendingIntent pIntent = PendingIntent.
                                getBroadcast(MainActivity.this, 0, intent, 0);
                                //获取系统闹钟服务
                                AlarmManager  aManager = (AlarmManager)
                                getSystemService(ALARM_SERVICE);
                                //设置服务在系统休眠时也会运行
                                aManager.set(AlarmManager.RTC_WAKEUP,
                                calendar.getTimeInMillis(), pIntent);
                                //显示设置的闹钟时间
                                String time = format(hourOfDay) + ":" +
                                format(minute);
                                textView.setText(time);
                                //Toast提示设置完成
                                Toast.makeText(MainActivity.this, "闹钟时
                                间为" + time, Toast.LENGTH_LONG).show();
                            }
                        },
                        hour,
                        minute,
                        true)
                    .show();
            }
        });
        //取消按钮监听
        cancle.setOnClickListener(new OnClickListener() {
            @Override
            public void onClick(View v) {
                // TODO Auto-generated method stub
                Intent intent = new Intent(MainActivity.this, AlarmReceiver.
                class);
                PendingIntent pIntent = PendingIntent.getBroadcast
```

```
                    (MainActivity.this, 0, intent, 0);
                //取消闹钟
                AlarmManager aManager = (AlarmManager)getSystemService
                (ALARM_SERVICE);
                aManager.cancel(pIntent);
                textView.setText("无闹钟设置");
                Toast.makeText(MainActivity.this, "无闹钟设置", Toast.
                LENGTH_LONG).show();
            }
        });
    }
    //使用两位数表示时间
    private String format(int x) {
    String s = "" + x;
    if(s.length() == 1)
        s = "0"+s;
    return s;
    }
}
```

（2）新建 AlarmActivity 显示闹钟对话框,代码如下。

```
public class AlarmActivity extends Activity {
    @Override
    protected void onCreate(Bundle savedInstanceState) {
        // TODO Auto-generated method stub
        super.onCreate(savedInstanceState);
        /创建一个对话框
        new AlertDialog.Builder(AlarmActivity.this)
        //对话框图标
        .setIcon(R.drawable.ic_launcher)
        //对话框标题
        .setTitle("时间到了! ")
        //对话框内容
        .setMessage("该起床了~")
        //对话框确定按钮监听
        .setPositiveButton("确定",
                new DialogInterface.OnClickListener() {
                    @Override
                    public void onClick(DialogInterface dialog, int which){
                        // TODO Auto-generated method stub
                        //单击"确定"按钮,结束当前Activity
                        finish();
                    }
                })
        //显示对话框
        .show();
    }
}
```

（3）新建 AlarmRecriver 监听系统时间,当到了闹铃时间时,弹出对话框提示,代码如下。

```
public class AlarmReceiver extends BroadcastReceiver {
    @Override
    public void onReceive(Context context, Intent intent) {
        // TODO Auto-generated method stub
```

```java
        //创建 Intent 对象
        Intent i = new Intent(context,AlarmActivity.class);
        //设置 Flag,将目标 Activity 压入栈顶
        i.setFlags(Intent.FLAG_ACTIVITY_NEW_TASK);
        //启动 Activity
        context.startActivity(i);
    }
}
```

(4) 在 AndroidManifest.xml 中注册使用 AlarmActivity 和 AlarmRecriver,代码如下。

```xml
        <!--注册 AlarmReceiver-->
        <receiver android:name="AlarmReceiver" android:process=":remote" />
         <!--注册 AlarmActivity-->
        <activity android:name="AlarmActivity"></activity>
```

第 17 章　Android 应用开发——网上购书

本章将介绍一个 Android 应用程序——网上购书。该程序只是模拟了网上购书的过程，并没有真正地联网实现网上购书功能。该案例的开发，将综合运用 Android 中的多种控件，以及数据库的操作等，帮助读者掌握实用的开发技巧。

17.1　系 统 简 介

本节在整体上介绍该应用程序的功能以及开发环境，使读者在进入正式的程序开发之前对系统有一定了解。

17.1.1　功能概述

该应用程序的主要功能包括以下几个部分，如图 17.1 所示。
- 用户登录；
- 用户浏览图书信息；
- 选择图书添加到购物车；
- 选择图书添加到收藏夹；
- 查看购物车或者收藏夹中的图书信息。

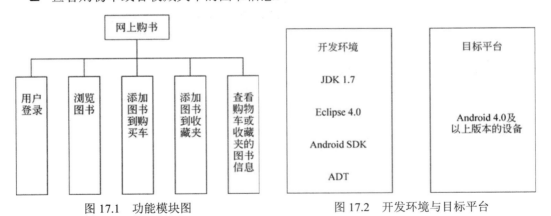

图 17.1　功能模块图　　　　　图 17.2　开发环境与目标平台

17.1.2　开发环境及目标平台

开发该系统需要用到的开发环境，以及目标平台，如图 17.2 所示。

17.2 系统架构

本节对系统的框架进行简要地介绍，以帮助读者更好地理解系统的开发过程。该系统可以分为 4 大模块，即登录模块、创建数据库数据表模块、图书列表模块和存储模块，如图 17.3 所示。

图 17.3 系统架构图

17.3 用户登录模块的实现

从本节开始，我们正式进入程序的代码开发。用户登录界面就是我们进入网上购书系统的入口，只有输入用户名和密码登录成功，才可以开始网上购书。登录界面如图 17.4 所示。

图 17.4 登录界面

用户登录界面需要通过填写用户名及密码，确认用户身份。只有当用户名及密码填写正确，才能进入网上购书系统。如果用户名或者密码填写不正确，则无法进入系统，Toast 提示"输入错误"。逻辑代码如下：

```
public class ShoppingActivity extends Activity {
    //输入用户名和密码
    private EditText edUser,edPassword;
```

```java
//登录按钮
private Button login;
//用户名和密码
private String user,password;
@Override
public void onCreate(Bundle savedInstanceState) {
    super.onCreate(savedInstanceState);
    setContentView(R.layout.activity_shopping);
    edUser = (EditText)findViewById(R.id.editText1);
    edPassword = (EditText)findViewById(R.id.editText2);
    login = (Button)findViewById(R.id.button1);
    login.setOnClickListener(new OnClickListener() {
        public void onClick(View v) {
            // TODO Auto-generated method stub
            //获取用户名和密码
            user = edUser.getText().toString();
            password = edPassword.getText().toString();
            if (user.equals("abc")&&password.equals("123")) {
                                //当用户名为"abc"时,密码为"123"
                Intent intent = new Intent();
                intent.setClass(ShoppingActivity.this,GoodsList.class);
                //单击"登录"按钮跳转到图书清单界面
                startActivity(intent);
            edUser.setText("");//用户名和密码,输入框置为空
                edPassword.setText("");
            }else {
                Toast.makeText(ShoppingActivity.this, "输入错误",
                Toast.LENGTH_LONG).show();
            //否则 Toast 提示输入错误
            }
        }
    });
}
}
```

17.4 数据库与数据表的实现

当用户在购书过程中,无论是将中意的书添加到购物车购买,还是将感兴趣的书添加到收藏夹收藏,都需要创建数据库来保存这些图书信息。我们创建了一个包含两张数据表的数据库,一张表用于保存购物车中的图书数据;另一张表用于保存收藏夹的图书数据,根据图书名称添加图书记录到对应数据表中。逻辑代码如下:

```java
public class ShoppingDatabaseAdapter {
    //购物车数据表的字段名称
public final static String KEY_NAME = "name";
    //购物车数据表的主键
    public static final String KEY_ROWID = "_id";
    //收藏夹数据表的字段名称
public static final String COLLECTION_NAME = "name";
    //收藏夹数据表的主键
    public final static String COLLECTION_ROWID = "_id";
    //上下文环境变量,传入构造方法作参数
    private final Context context;
    //数据库辅助类子类变量
```

```java
    private DatabaseOpenHelper dHelper;
    //数据库变量
    private SQLiteDatabase sDatabase;
    //数据库名称
    private static final String DATABASE_NAME = "data.db";
    //购物车数据表名称
    private static final String STORE_TABLE = "store";
    //收藏夹数据表名称
    private static final String COLLECTION_TABLE = "collection";
    //数据库版本
    private static final int DATABASE_VERSION = 1;
    //创建购物车数据表语句
    private static final String DATABASE_CREATED01 = "create table " +
            STORE_TABLE + " (" + KEY_ROWID +
            " integer primary key autoincrement, " +
            KEY_NAME + " text not null) " ;
    //创建收藏夹数据表语句
    private static final String DATABASE_CREATED02 = "create table " +
            COLLECTION_TABLE + "(" + COLLECTION_ROWID +
            " integer primary key autoincrement, " +
            COLLECTION_NAME + " text not null) ";
    //继承于数据库辅助类
    private class DatabaseOpenHelper extends SQLiteOpenHelper{
        public DatabaseOpenHelper(Context context, String name,
        CursorFactory factory, int version) {

            super(context, DATABASE_NAME, factory, DATABASE_VERSION);
        }
        @Override
        public void onCreate(SQLiteDatabase db) {
            // TODO Auto-generated method stub
            //执行创建购物车数据表和收藏夹数据表的语句
            db.execSQL(DATABASE_CREATED01);
            db.execSQL(DATABASE_CREATED02);
        }
        @Override
        public void onUpgrade(SQLiteDatabase db, int oldVersion, int newVersion){
            // TODO Auto-generated method stub
            db.execSQL("drop table if exists data.db");  //执行删除表语句
        }
    }
    public ShoppingDatabaseAdapter(Context ctx) {        //构造方法
        this.context = ctx;
    }
public ShoppingDatabaseAdapter open() throws SQLException {
                                                       //自定义 open()方法
        dHelper = new DatabaseOpenHelper(context, null, null, 0);
                                                  //获取可读可写的数据库
        sDatabase = dHelper.getWritableDatabase();
         return this;                             //返回可读可写的数据库对象
    }
    public void close() {                             //关闭数据库
        dHelper.close();
    }
    public long store_createNote(String name) {    //根据图书名称存入购物车
        ContentValues values = new ContentValues();
        values.put(KEY_NAME, name);
        return sDatabase.insert(STORE_TABLE, null, values);
    }
    public long collection_createNote(String name) {
                                                   //根据图书名称存入收藏夹
    ContentValues values = new ContentValues();
```

```
        values.put(COLLECTION NAME, name);
        sDatabase.insert(COLLECTION TABLE, null, values);
        return 0 ;
    }
    public Cursor store_fetchAllNotes() {        //自定义方法查询购物车数据
        return sDatabase.query(STORE TABLE, new String[] {KEY ROWID,
            KEY NAME}, null, null, null, null, null);
    }
    public Cursor collection_fetchAllNotes(){    //自定义方法查询收藏夹数据
        return         sDatabase.query(COLLECTION TABLE,    new    String[]
{COLLECTION ROWID,
            COLLECTION NAME}, null, null, null, null, null);
    }
    public void drop_Dababase(){                 //删除数据表
        sDatabase.execSQL("drop table store if exists data.db");
        sDatabase.execSQL("drop table collection if exists data.db");
    }
}
```

17.5　图书浏览选择模块的实现

用户登录成功后，进入网上购书系统，就可以浏览图书了。单击图书，界面弹出文本框，显示图书的单价、作者、出版社等详细信息供读者参考，如图 17.5 所示。用户也可以选择中意的图书，添加到购物车准备购买；或者选择感兴趣图书，添加到收藏夹继续关注，如图 17.6 所示。

图 17.5　图书详细信息

图 17.6　选择添加图书

逻辑代码如下：

```java
public class GoodsList extends Activity {
    //图书按钮
    private ImageButton imageButton1,imageButton2,imageButton3;
    //选择按钮
    private CheckBox checkBox1,checkBox2,checkBox3;
    //"添加到购物车"和"添加到收藏夹"按钮
    private Button button1,button2;
    //我的购物车、我的收藏夹、退出文本
    private TextView textView1,textView2,textView3;
    //收藏夹标志,flag 值为 true 时表示购物车
    public static boolean flag = false;
    // ShoppingDatabaseAdapter 类变量
    ShoppingDatabaseAdapter sDatabaseAdapter;
        public void onCreate(Bundle savedInstanceState) {
            super.onCreate(savedInstanceState);
            setContentView(R.layout.goods_list);
            //调用构造方法创建 ShoppingDatabaseAdapter 类对象
            sDatabaseAdapter = new ShoppingDatabaseAdapter(this);
            //调用 open()方法获取数据库
            sDatabaseAdapter.open();
            checkBox1 = (CheckBox)findViewById(R.id.checkBox1);
            checkBox2 = (CheckBox)findViewById(R.id.checkBox2);
            checkBox3 = (CheckBox)findViewById(R.id.checkBox3);
            imageButton1 = (ImageButton)findViewById(R.id.imageButton1);
            //单击图书按钮,弹出对话框显示对应图书信息
            imageButton1.setOnClickListener(new OnClickListener() {
                public void onClick(View v) {
                    // TODO Auto-generated method stub
                    AlertDialog.Builder builder=new AlertDialog.Builder
                    (GoodsList.this);
                    builder.setIcon(R.drawable.android)
                            .setTitle("Android 开发入门与实战体验")
                    .setMessage("单价：69.80\n 作者：李佐彬 编著\n 出版社：机械
                    工业出版社");
                    builder.show();
                }
            });

            imageButton2 = (ImageButton)findViewById(R.id.imageButton2);
            //单击图书按钮,弹出对话框显示对应图书信息
            imageButton2.setOnClickListener(new OnClickListener() {
                public void onClick(View v) {
                    // TODO Auto-generated method stub
                    AlertDialog.Builder builder=new AlertDialog.Builder
                    (GoodsList.this);
                    builder.setIcon(R.drawable.java)
                            .setTitle("零基础学 java")
                            .setMessage("单价：40.00\n 作者：王鹏 著\n 出版社：
                            机械工业出版社");
                    builder.show();
                }
            });

            imageButton3 = (ImageButton)findViewById(R.id.imageButton3);
            //单击图书按钮,弹出对话框显示对应图书信息
```

```java
imageButton3.setOnClickListener(new OnClickListener() {
    public void onClick(View v) {
        // TODO Auto-generated method stub
        AlertDialog.Builder builder=new AlertDialog.Builder(GoodsList.this);
        builder.setIcon(R.drawable.c)
                .setTitle("C#程序设计语言")
                .setMessage("单价：32.00\n作者：杜松江 著\n出版社：北京邮电大学出版社");
        builder.show();
    }
});

button1 = (Button)findViewById(R.id.button1);
//添加到购物车按钮监听
button1.setOnClickListener(new OnClickListener() {
    public void onClick(View v) {
        // TODO Auto-generated method stub
        if (v==button1) {
            //当选中第一个选择按钮
            if (checkBox1.isChecked()) {
                //在购物车数据表中添加Android图书记录
                sDatabaseAdapter.store_createNote("Android开发入门与实战体验");
            }
            //当选中第二个选择按钮
            if (checkBox2.isChecked()) {
                //在购物车数据表中添加Java图书记录
                sDatabaseAdapter.store_createNote("零基础学java");
            }

            //当选中第三个选择按钮
            if (checkBox3.isChecked()) {
                //在购物车数据表中添加C#图书记录
                sDatabaseAdapter.store_createNote("C#程序设计语言");
            }
        }
    }
});
button2 = (Button)findViewById(R.id.button2);
//添加到收藏夹按钮监听
button2.setOnClickListener(new OnClickListener() {
    public void onClick(View v) {
        // TODO Auto-generated method stub
        if (v==button2) {
            //当选中第一个选择按钮
            if (checkBox1.isChecked()) {
                //在收藏夹数据表中添加Android图书记录
                sDatabaseAdapter.collection_createNote
                    ("Android开发入门与实战体验");
            }
            //当选中第二个选择按钮
            if (checkBox2.isChecked()) {
                //在收藏夹数据表中添加Java图书记录
                sDatabaseAdapter.collection_createNote("零基础学java");
            }
            //当选中第三个选择按钮
```

```java
                if (checkBox3.isChecked()) {
                    //在收藏夹数据表中添加C#图书记录
                    sDatabaseAdapter.collection_createNote("C#程
                        序设计语言");
                }
            }
        }
    });
    textView1 = (TextView)findViewById(R.id.textView1);
    //我的购物车文本监听
    textView1.setOnClickListener(new OnClickListener() {

        public void onClick(View v) {
            // TODO Auto-generated method stub
            flag=true;
            //单击跳转到存储界面显示购物车中图书信息
            Intent intent=new Intent(GoodsList.this, Shop_store.class);
            //启动目标Activity
            startActivity(intent);
        }
    });
    textView2 = (TextView)findViewById(R.id.textView2);
    //我的收藏夹文本监听
    textView2.setOnClickListener(new OnClickListener() {
        public void onClick(View v) {
            // TODO Auto-generated method stub
            //收藏夹标志
            flag=false;
            //单击跳转到存储界面显示收藏夹中图书信息
            Intent intent=new Intent(GoodsList.this, Shop_store.class);
            //启动目标Activity
            startActivity(intent);
        }
    });
    textView3 = (TextView)findViewById(R.id.textView3);
    //退出文本监听
    textView3.setOnClickListener(new OnClickListener() {

        public void onClick(View v) {
            // TODO Auto-generated method stub
            //单击退出文本,结束当前Activity,退出购书系统
            finish();
        }
    });
}
}
```

17.6 存储模块的实现

当用户添加图书到购物车,或者添加图书到收藏夹,系统都会在对应的数据表中插入相应的图书数据记录,然后用户可以单击界面中的"我的购物车"和"我的收藏夹"按钮,查看图书信息。如图17.7所示,购物车中有一条图书记录,收藏夹中有两条图书记录。

图 17.7 查看图书存储界面

逻辑代码如下：

```java
@TargetApi(11)
public class Shop_store extends Activity {
    //显示欢迎字幕
    private TextView tvWelcome;
    //返回按钮
    private Button back;
    //显示图书记录
    private ListView listView;
    //数据库
    ShoppingDatabaseAdapter sDatabaseAdapter;
    @Override
    protected void onCreate(Bundle savedInstanceState) {
        // TODO Auto-generated method stub
        super.onCreate(savedInstanceState);
        setContentView(R.layout.store_list);
        //创建 ShoppingDatabaseAdapter 对象
        sDatabaseAdapter = new ShoppingDatabaseAdapter(this);
        //创建数据库
        sDatabaseAdapter.open();
        tvWelcome = (TextView)findViewById(R.id.textView1);
        listView = (ListView)findViewById(R.id.listView1);
        back = (Button)findViewById(R.id.button1);
        back.setOnClickListener(new Button.OnClickListener() {
            public void onClick(View v) {
                //结束当前 Activity,返回到上一界面
                finish();
            }
        });
        //查看图书存储录
        renderList();
    }
    //自定义方法查看图书存储录
    public void renderList(){
        //我的购物车
        if (GoodsList.flag) {
            //重置欢迎字幕
            tvWelcome.setText("欢迎您,abc! 您的购物车内有以下物品: ");
```

```
            //查看购物车数据,返回 Cursor 对象,作适配器参数
            Cursor cursor = sDatabaseAdapter.store_fetchAllNotes();
            //创建适配器,为 ListView 提供图书数据
            SimpleCursorAdapter sAdapter = new SimpleCursorAdapter(this,
            R.layout.store_row , cursor, new String[]
            {ShoppingDatabaseAdapter.KEY_NAME,
            ShoppingDatabaseAdapter.COLLECTION_ROWID},
            new int[] {R.id.textView1,R.id.textView2},1);
                //为 ListView 绑定适配器
                listView.setAdapter(sAdapter);
}
        //我的收藏夹
        if (!GoodsList.flag) {
        //重置欢迎字幕
            tvWelcome.setText("欢迎您,abc! 您的收藏夹内有以下物品: ");
            //查看收藏夹数据,返回 Cursor 对象,作适配器参数
            Cursor cursor = sDatabaseAdapter.collection_fetchAllNotes();
            //创建适配器,为 ListView,提供图书数据
            SimpleCursorAdapter sAdapter = new SimpleCursorAdapter
             (this,R.layout.store_row ,
                    cursor, new String[]
                    {ShoppingDatabaseAdapter.KEY_NAME,
                    ShoppingDatabaseAdapter.COLLECTION_ROWID},
                    new int[] {R.id.textView1,R.id.textView2},2);
            //为 ListView 绑定适配器
            listView.setAdapter(sAdapter);
        }
    }
}
```

17.7 小　　结

本章以模拟网上购书的开发为例,涉猎 Android 中的诸多开发技巧。希望读者可以参考本章中的核心代码,基于本案例,开发出真正可以实现网上购书的应用程序。更进一步地掌握 Android 项目的开发,达到学以致用的目的。